MEMOIRS
of the
American Mathematical Society

W0115375

Number 449

A Classification Theorem
for Homotopy Commutative H-Spaces
with Finitely Generated
mod 2 Cohomology Rings

Michael Slack

July 1991 • Volume 92 • Number 449 (second of 4 numbers) • ISSN 0065-9266

American Mathematical Society
Providence, Rhode Island

1980 *Mathematics Subject Classification* (1985 *Revision*).
Primary 55P45, 55S12, 55S20, 55S35, 55S45;
Secondary 55P15, 55P47, 55P60, 55R05.

Library of Congress Cataloging-in-Publication Data

Slack, Michael, 1963–
  A classification theorem for homotopy commutative H-spaces with finitely generated mod 2 cohomology rings/Michael Slack.
    p. cm. – (Memoirs of the American Mathematical Society, ISSN 0065-9266; no. 449)
    Includes bibliographical references.
    ISBN 0-8218-2514-3
    1. H-spaces. 2. Obstruction theory. 3. Dyer-Lashof operations. I. Title. II. Series.
QA3.A57 no. 449
[QA612.77]
510 s–dc20   91-13084
[514′.24]   CIP

Subscriptions and orders for publications of the American Mathematical Society should be addressed to American Mathematical Society, Box 1571, Annex Station, Providence, RI 02901-1571. *All orders must be accompanied by payment.* Other correspondence should be addressed to Box 6248, Providence, RI 02940-6248.

SUBSCRIPTION INFORMATION. The 1991 subscription begins with Number 438 and consists of six mailings, each containing one or more numbers. Subscription prices for 1991 are $270 list, $216 institutional member. A late charge of 10% of the subscription price will be imposed on orders received from nonmembers after January 1 of the subscription year. Subscribers outside the United States and India must pay a postage surcharge of $25; subscribers in India must pay a postage surcharge of $43. Expedited delivery to destinations in North America $30; elsewhere $82. Each number may be ordered separately; *please specify number* when ordering an individual number. For prices and titles of recently released numbers, see the New Publications sections of the NOTICES of the American Mathematical Society.

BACK NUMBER INFORMATION. For back issues see the AMS Catalogue of Publications.

MEMOIRS of the American Mathematical Society (ISSN 0065-9266) is published bimonthly (each volume consisting usually of more than one number) by the American Mathematical Society at 201 Charles Street, Providence, Rhode Island 02904-2213. Second Class postage paid at Providence, Rhode Island 02940-6248. Postmaster: Send address changes to Memoirs of the American Mathematical Society, American Mathematical Society, Box 6248, Providence, RI 02940-6248.

Table of Contents

Abstract

It is shown that a 2–connected homotopy commutative H-space with associative mod 2 homology ring and finitely generated mod 2 cohomology ring has acyclic mod 2 cohomology. This implies that a connected, homotopy commutative, homotopy associative H-space with finitely generated mod 2 cohomology ring is mod 2 homotopy equivalent to a product of Eilenberg–MacLane spaces, giving a complete classification of such spaces localized at the prime 2.

Key words and phrases:

H-space, homotopy commutative, loop space, Steenrod algebra, secondary operations, Hopf algebra.

Received by the editor October 27, 1989 and in revised form January 15, 1991.

§0 Introduction

A. Statement of results

Unless stated otherwise, all spaces will be localized at the prime 2, path connected and have the homotopy type of a CW complex, and all coefficients will be Z_2. Let X be a homotopy commutative H-space whose mod 2 cohomology is coassociative and finitely generated as an algebra. The main objective of this paper is to give a proof of the following classification theorem.

THEOREM 0.1 (MAIN THEOREM).   *If X is 2-connected, then the mod 2 cohomology of X is acyclic.*

Theorem 0.1 admits the following corollaries.

COROLLARY 0.2.   *If X is homotopy associative, then X is mod 2 homotopy equivalent to a product of Eilenberg–MacLane spaces from the list ($r \geq 1$):*

$$K(Z,1), \ K(Z_{2^r},1), \ K(Z,2). \tag{0.1}$$

Research partially supported by a Sloan Doctoral Dissertation Fellowship.

1

In particular, the mod 2 cohomology of X is primitively generated.

COROLLARY 0.3.   *If X is simply connected, then the integral cohomology of X is two*

torsion free.

There are many known examples of homotopy associative H-spaces with mod 2 coho-

mology finitely generated as an algebra. For this discussion, spaces will not be localized.

If Y is a simply connected finite loop space (e.g. a compact simple Lie group), then Clark

[3] has shown that $\pi_3(Y) \neq 0$ (it is not known if this holds at the prime 2). We may form

the fibration

$$
\begin{array}{c}
Y\langle 3\rangle \\
\downarrow \\
Y \xrightarrow{\ f\ } K(\pi_3(Y),3)
\end{array}
\qquad (0.2)
$$

The fibre $Y\langle 3\rangle$ is called the 3-connective cover of Y. A property of $Y\langle 3\rangle$ is that its mod

2 cohomology is finitely generated as an algebra. This follows by applying the Eilenberg–

Moore spectral sequence to the fibration above. Suppose now that $\{Y_i\}_{i\in S}$ is a set of

simply connected finite loop spaces. Let Q and R be finite subsets of the indexing set

S such that their union is non-empty. Set $W = \times_{j\in Q}(Y_j\langle 3\rangle) \times_{k\in R} (Y_k)$; this is a loop

space with finitely generated mod 2 cohomology. Applying the Main Theorem yields the

following corollary.

COROLLARY 0.4.   *There is no choice of multiplication on W which is homotopy commu-*

tative.

The motivation for the Main Theorem comes from the following well known theorem.

THEOREM 0.5.   *If X is a homotopy commutative H-space which is a finite complex, then X has the homotopy type of a torus.*

B. The torus theorem

Theorem 0.5 is one of the oldest and most well known theorems in the study of H-spaces, known as the torus theorem, due originally to Hubbuck [6]. Hubbuck's proof uses K-theory techniques, and relies on a result of Browder [1] which states that a homotopy commutative finite H-space has cohomology which is an exterior algebra on odd dimensional generators and is torsion free. Hubbuck proves that all the generators must lie in cohomological dimension one. This implies that the universal cover of X must be acyclic, which is enough to show that X has the homotopy type of a torus. Along the same lines, Lin [10] gives the following generalization of the torus theorem.

THEOREM 0.6.   *If X is a simply connected, homotopy commutative H-space whose mod 2 cohomology is an exterior algebra on odd degree generators and $PH^{odd}(X)$ is finite dimensional, then $H^*(X)$ is acyclic.*

One of the striking features of this theorem is that it shows that the torus theorem

depends on the mod 2 homotopy type of X. Hubbuck's proof also depends on the mod 2

structure.

Theorem 0.1 is an attempt to generalize the torus theorem to non-finite, homotopy

commutative H-spaces whose mod 2 cohomology is finitely generated as an algebra. Un-

fortunately, the line of proof requires some associativity hypothesis. However, in the case

of a finite H-space, work of Browder [1] implies that the mod 2 homology is associative

and torsion free and that the mod 2 cohomology is primitively generated. It also shows

that a 1-connected finite homotopy commutative H-space is 2-connected. Therefore, in

combination with Browder's results, Theorem 0.1 implies that any 1-connected, finite ho-

motopy commutative H-space is acyclic. In this sense, the proof of the Main Theorem is

a significant generalization of Lin's cohomological proof of the torus theorem.

Iriye and Kono [7] have shown that any H-space localized at an odd prime has a

homotopy commutative multiplication. So in order to put a restriction on p-localized

H-spaces for odd primes, a stronger condition than homotopy commutativity is needed.

Such a condition is given by Hemmi [5]. He shows that if a (p-localized) space is simply

connected with finite mod p cohomology, and is what he calls a quasi C_p-space, then

its mod p cohomology is acyclic. In the case where $p = 2$, a quasi C_2-space corresponds precisely to a homotopy commutative H-space. McGibbon [13] has proven a similar result. In lieu of the results given in this work, one might conjecture that when p is an odd prime, a p-local quasi C_p-space with finitely generated mod p cohomology is mod p equivalent to a product of Eilenberg-MacLane spaces.

C. Acknowledgments

The work in this paper is comprised of a portion of the author's dissertation completed at UC San Diego. I am very grateful to my friend and advisor, Jim Lin, for his patience and many hours of help. I also would like to thank Frank Williams for his help with the h_n-deviation, and Michael Freedman for his generous support. Jim and Frank also deserve thanks for proving Proposition 5.7, which is essential in the proof of Theorem 0.1 and will appear separately [8].

§1 Techniques used in the proof

The proof of the Main Theorem consists of several steps. Most of the steps in the proof involve secondary operation arguments. The reader who is unfamiliar with these techniques is recommended to refer to [9], [10] and [15]. The proof consists of a systematic study of QH^*, the module of indecomposables of H^*. In section 2, the proof embarks with a study of QH^{even}. In that section it is shown that the only indecomposables of even degree must lie in degrees which are powers of two. Having cleared away most of QH^* in even degrees, the study of QH^{odd} begins in section 3. In section 3 it is shown that there are no indecomposables in degrees $2^r(2k+1) - 1$ for $r \geq 2$ and $k \geq 1$, and when $r = 1$ and k is not a power of two. This means that the only odd degree indecomposables remaining at the end of section 3 are in degrees which are a power of two plus or minus one. In section 4 we show that there are no indecomposables in degrees greater than 9, which leaves QH^* possibly non-zero in degrees 3, 4, 5, 7, 8 and 9. Then, in section 5, it is shown that $QH^n = 0$ for $n = 3$, 7, 8 and 9. Also included there is a statement of a theorem of Lin and Williams [11], which essentially states that there are no indecomposables in degrees

4 and 5. This completes the proof of the Main Theorem. The proofs of the corollaries are given in section 6, and section 7 (the appendix) contains a list of notation conventions.

There are several techniques used in the proof of the Main Theorem and its corollaries, some old and some new. However, many of the new techniques are variants of old themes. An "implication"argument of Browder [1] is generalized in section 2 to show that Sq^1 maps certain odd degree primitives to fourth power elements. Another variant on an old theme is the factorization of $Sq^8 Sq^4$ through primary and secondary operations. This is used in several places; and it should be pointed out that the existence of such factorizations was originally drawn to my attention by Jim Lin [12]. In fact, such a factorization is used in the proof of Proposition 5.7 by Lin and Williams [11]. Probably the biggest new contribution to the proof of the Main Theorem is the h_n-deviation, which is used extensively.

We give a brief review of the definition and properties of the h_n-deviation; the proofs will be assumed and can be found in a forthcoming paper [15]. A space X is called an H_n-space if there is a Z_2-equivariant map $\theta_n^X : S^n \times X^2 \to X$ such that $\theta_n^X(w; *, x) = \theta_n^X(w; x, *) = x$. The action of the nontrivial element of Z_2 on $S^n \times X^2$ is $(w; x_1, x_2) \to (a(w); x_2, x_1)$, where $a : S^n \to S^n$ is the antipodal map; and Z_2 acts trivially on X. Suppose we are given H_n-spaces X and K, and a map $f : X \to K$. Suppose also that the diagram below commutes up to Z_2-equivariant homotopy.

$$
\begin{array}{ccc}
S^{n-1} \times X^2 & \xrightarrow{\theta^X_{n-1}} & X \\
\downarrow{\scriptstyle 1 \times f^2} & {\scriptstyle \theta^K_{n-1}} & \downarrow{\scriptstyle f} \\
S^{n-1} \times K^2 & \xrightarrow{\theta^K_{n-1}} & K
\end{array}
\tag{1.1}
$$

We say here that f is an h_{n-1}-map. (Note that an H_n-space is an H_{n-1}-space by restricting

θ_n the the equator of S^n). The h_n-deviation is then the obstruction to extending the

diagram above to a commutative (up to Z_2-equivariant homotopy) diagram

$$
\begin{array}{ccc}
S^n \times X^2 & \xrightarrow{\theta^X_n} & X \\
\downarrow{\scriptstyle 1 \times f^2} & {\scriptstyle \theta^K_n} & \downarrow{\scriptstyle f} \\
S^n \times K^2 & \xrightarrow{\theta^K_n} & K
\end{array}
\tag{1.2}
$$

If $f : X \to K$ is an h_{n-1}-map between H_n-spaces X and K, then the h_n-deviation of f,

$h_n(f)$, lies in the group $[X \wedge X, \Omega^n K]$. In the cases that $n = 0$ or $n = 1$, this obstruction

is known, respectively, as the H-deviation or c-deviation. In the special case that K is an

Eilenberg-MacLane space, $h_n(f)$ is a cohomology class. This cohomology class is related

to the Dyer-Lashof operation Q_n by the following theorem [15].

THEOREM 1.1.   *Given an h_{n-1}-map $f : X \to K$ between H_n-spaces (for $n \geq 1$), there is*

an element $h_n(f)$ in $[X \wedge X, \Omega^n K]$ called the h_n-deviation of f which vanishes if and only

if f is an h_n-map, and satisfies the following properties:

(A)   If $K = K(Z_2, r)$ so that f represents a cohomology class x, then there is a homology

class $\overline{y}$ such that $\langle \overline{y} \otimes \overline{y}, h_n(f) \rangle \neq 0$ if and only if $Q_n(\overline{y}) = Q^{\frac{r+n}{2}}(\overline{y}) = \overline{x}$, where $\overline{x}$ is a

homology dual to x, and $\langle \ , \ \rangle$ is the Kronecker pairing.

(B) If E is the second stage of a Postnikov system with k-invariant given by the squaring map $\xi : K(Z_2, r) \to K(Z_2, 2r)$ (i.e. $\xi^*(\iota_{2r}) = \iota_r^2$), and if $r > n + 2$, then $\Omega^{n+1}E \simeq$

$K(Z_2, r-n-1) \times K(Z_2, 2r-n-2)$ as H_{n-1}-spaces and $[h_n(\iota_{2r-n-2})] = \iota_{r-n-1} \otimes \iota_{r-n-1}$

(i.e. $Q_n(\bar{\iota}_{r-n-1}) \neq 0$).

(C) $h_{n+1}(\Omega f)$ is homotopic to the adjoint of the composition

$$\Sigma\Omega X \wedge \Sigma\Omega X \xrightarrow{\epsilon \wedge \epsilon} X \wedge X \xrightarrow{h_n(f)} \Omega^n K \tag{1.3}$$

where ϵ is the evaluation map, i.e. if (λ, r) is a Moore loop, then $\epsilon(t, (\lambda, r)) = \lambda(rt)$.

In particular, if K is an Eilenberg-MacLane space, then $[h_{n+1}(\Omega f)] = (\sigma^* \otimes \sigma^*)[h_n(f)]$

(corresponding to the fact that $\sigma_* Q_{n+1} = Q_n \sigma_*$), where σ_* (resp. σ^*) is the homology

(resp. cohomology) suspension homomorphism.

(D) If $f : X \to Y$ and $g : Y \to Z$ are h_{n-1}-maps between H_n-spaces, then gf is an

h_{n-1}-map and $h_n(gf) : X \wedge X \to \Omega^n Z$ is homotopic to

$$h_n(g)(f \wedge f) + (\Omega^n g)h_n(f) \tag{1.4}$$

(E) If X and K are H_{n+1}-spaces and $f : X \to K$ is an h_{n-1}-map, then $[h_n(f)] \in$

$\ker(1 + (-1)^{n+1}T^*)$, where T is the switching map $X \wedge X \to X \wedge X$.

We will give a brief review of the use of the h_n-deviation in secondary operation

arguments. Suppose that X is an H_n-space and $x \in H^k$ is such that the map $f : X \to$

$K(Z_2, k)$ representing x is an h_{n-1}-map. Suppose further that there is a relation in the

Steenrod algebra $\mathcal{A}(2)$ given by $Sq^m = \sum_i \alpha_i \beta_i$, where $m = k + n + 1$, and $\beta_i x = 0$ for all

i. Then there is a commutative diagram (of fibration towers) of H_n-spaces

$$
\begin{array}{ccc}
\prod_i K(Z_2, \deg \beta_i + k - 1) & \xrightarrow{\Omega h} & K(Z_2, 2k + n) \\
\downarrow j & & \downarrow \\
\tilde{f} \nearrow \quad E_0 & \xrightarrow{v} & E_1 \simeq K(Z_2, k) \times K(Z_2, 2k + n) \\
\downarrow p & & \downarrow \\
X \xrightarrow{f} K_0 = K(Z_2, k) & \xrightarrow{id} & K_0 \\
\downarrow g & & \downarrow Sq^m \\
K_1 = \prod_i K(Z_2, \deg \beta_i + k) & \xrightarrow{h} & K(Z_2, 2k + n + 1)
\end{array}
\tag{1.5}
$$

where g^* is the row matrix (β_i) and h^* is the column matrix (α_i). The element $\iota_{2k+n} \in$

$H^*(E_1)$ has h_n-deviation $h_n(\iota_{2k+n}) = \iota_k \otimes \iota_k$. Although it is not necessarily the case,

assume that the lifting $\tilde{f}$ can be chosen to be an h_{n-1}-map. Then $h_n(v \circ \tilde{f}) = h_n(v)(\tilde{f} \wedge$

$\tilde{f}) + (\Omega^n v) h_n(\tilde{f})$ and thus

$$
(h_n(v \circ \tilde{f}))^*(\iota_{2k+n}) = (\tilde{f} \wedge \tilde{f})^* h_n(v)^*(\iota_{2k+n}) + h_n(\tilde{f})^*(\Omega^n v)^*(\iota_{2k+n}).
\tag{1.6}
$$

Set $\phi(x) = (v \circ \tilde{f})^*(\iota_{2k+n})$. Then $h_n(\phi(x)) = (\tilde{f}^* \otimes \tilde{f}^*)(v^* \iota_k \otimes v^* \iota_k) + h_n(\tilde{f})^*((\sigma^*)^n v^* \iota_{2k+n})$

so that

$$
h_n(\phi(x)) = x \otimes x + h_n(\tilde{f})^* w,
\tag{1.7}
$$

where $w \in H^{2k}(E_0)$ is equal to $(\sigma^*)^n v^* \iota_{2k+n}$. If $h_n(f) = h_n(x) \in B \otimes B$, where B is an

$\mathcal{A}(2)$ invariant sub-Hopf algebra of $H^*(X)$, then an application of the Cartan formula for secondary operations (see Thomas [16]) yields

$$h_n(\tilde{f})^* w \in H^* \otimes B + B \otimes H^* + \operatorname{im} \sum_i \alpha_i \tag{1.8}$$

Combining (1.5) and (1.6) gives the formula

$$h_n(\phi(x)) = x \otimes x + (H^* \otimes B + B \otimes H^* + \operatorname{im} \sum_i \alpha_i + \operatorname{im}(1 + T^*)). \tag{1.9}$$

The $\operatorname{im}(1 + T^*)$ term comes from the indeterminacy in the definition of the h_n-deviation.

If $x \otimes x$ is not contained in the total indeterminacy, then this operation detects a non-zero

$2k + n$ dimensional h_{n-1} class in H^* (or equivalently, a non-zero Dyer-Lashof operation

Q_n in H_*).

One difficulty in applying this operation is getting the lifting $\tilde{f}$ to be an h_{n-1}-map.

The following technique is useful. If f is an h_{n-1}-map as assumed, and if $\tilde{f}$ is an h_{m-1}-

map for some $m < n$, then $h_m(\tilde{f})$ is defined and $h_m(f) = 0$. Thus $h_m(\tilde{f})$ factors as a

composition

$$X \wedge X \xrightarrow{h_m} \Omega^{m+1} K_1 \xrightarrow{\Omega^m j} \Omega^m E_0. \tag{1.10}$$

Under these circumstances there is a formula

$$(\Omega^{m+1} g) h_{m+1}(f) = (1 + (-1)^{m+1} T^*)[h_m]. \tag{1.11}$$

§2 Initial study of QH^{even}

The proof of the Main Theorem (Theorem 0.1) begins in this section. As suggested in the title of the section, the analysis begins with a study of the even degree indecomposables. Throughout this section X will be a space satisfying the hypotheses of Theorem 0.1. The main result of this section (Proposition 2.6) is that the only even degree indecomposables of H^* must lie in degrees which are powers of two.

The cohomology of X is a bicommutative, biassociative Hopf algebra. In their paper [14], Milnor and Moore prove that the following sequence is exact

$$0 \to P(\xi H^*) \to PH^* \to QH^* \to Q(\lambda H^*) \to 0, \qquad (2.1)$$

where the Hopf algebra λH^* is the dual of ξH_*. Let x be an indecomposable element of H^*. The projection of x to QH^* will be denoted by $\hat{x}$. The following definition will be useful in characterizing indecomposables.

DEFINITION 2.1.   *If x is an indecomposable element of H^* and the image of $\hat{x}$ under the map $QH^* \to Q(\lambda H^*)$ is non-zero, then x is called a **non-primitive generator**.*

12

Note that by the exact sequence (2.1), x is a non-primitive generator if and only if there

is no primitive element that projects to $\hat{x}$. Note also that every non-primitive generator

has even degree since λH^* is zero in odd degrees.

The main result of this section is Proposition 2.6, which says that QH^{even} is concen-

trated in degrees which are powers of two. The proof follows from three lemmas. The

first of these is Lemma 2.2, which gives a small amount of control on the action of Sq^1

applied to odd degree primitives. The knowledge of the action of Sq^1 is a crucial first step

in being able to apply secondary operation arguments. With Lemma 2.2 in hand it is pos-

sible to prove Lemma 2.4, which, roughly speaking, states that above the highest degree

(not counting degrees which are powers of two) in which there is a non-primitive generator

there are no even degree generators. This is a necessary step in the proof of Lemma 2.5,

which states that there are no non-primitive generators in degrees which are not powers

of two. This is shown inductively, by successsively showing that the highest degree not

equal to a power of two in which there is a non-primitive generator, is congruent to 0 mod

2^r for each $r \geq 1$. The fact that the mod 2 cohomology is finitely generated implies that

there can be no such generators. Then given that there are no non-primitive generators

in degrees which are not powers of two, Lemma 2.4 implies Proposition 2.6. The proofs of

Lemmas 2.4 and 2.5 are very similar, and use the method of secondary operations (ladder

Toda brackets) as can be found in Lin [9].

LEMMA 2.2. *If the largest degree not equal to a power of two such that there is a non-primitive generator is equal to n, and if $x \in PH^{odd}$ is such that $\deg x + 1$ is not a power of two and $\deg x > n$, then $Sq^1 x \in \xi^2 H^*$.*

PROOF: Pick the largest k not equal to a power of two such that there is an element $x \in PH^{2k-1}$ with $Sq^1 x \notin \xi^2 H^*$. Assume that the lemma is false, i.e. $2k - 1 > n$. Let $y = Sq^1 x$. Then $y \in PH^*$. There are two cases; either $y^2 = 0$ or $y^2 \neq 0$.

CASE 1: $y^2 = 0$. Consider the Postnikov system below.

$$K(Z_2, 4k - 1)$$

$$\downarrow j$$

$$E \xrightarrow{\ v\ } K(Z_2, 4k) \qquad\qquad (2.2)$$

$$\tilde{f} \nearrow \quad \downarrow p$$

$$X \xrightarrow{\ f\ } K(Z_2, 2k) \xrightarrow{Sq^{2k}} K(Z_2, 4k)$$

In the diagram, f represents the element y, E is the second stage of a Postnikov system with k-invariant Sq^{2k}, and v represents the element with the properties that $j^* v = Sq^1 \iota_{4k-1}$ and $\overline{\Delta}(v) = p^* \iota_{2k} \otimes p^* \iota_{2k}$ corresponding to the relation $Sq^{2k+1} = Sq^1 Sq^{2k}$. The map f lifts to $\tilde{f}$ since $Sq^{2k} y = y^2 = 0$. The projection p is an H-map, so the H-deviation of f

satisfies $D_f = pD_{\tilde{f}}$. Because y is primitive, the H-deviation of f is zero, so the H-deviation of $\tilde{f}$ factors as a composition

$$X \wedge X \xrightarrow{D} K(Z_2, 4k-1) \xrightarrow{j} E, \tag{2.3}$$

and the cohomology class $[D]$ satisfies the formula

$$(1 + T^*)[D] = Sq^{2k}h_1(f). \tag{2.4}$$

But $h_1(f) \in H^{2k-1}(X \wedge X)$, so $Sq^{2k}h_1(f) = 0$. Under these circumstances, there is a secondary operation ϕ and a formula

$$\overline{\Delta}\phi(y) = y \otimes y + Sq^1[D], \tag{2.5}$$

and $[D] \in \ker(1 + T^*)$. By the Cartan formula, Sq^1 applied to elements in $\ker(1 + T^*)$ gives elements in $\text{im}(1 + T^*)$. Thus

$$\overline{\Delta}\phi(y) = y \otimes y + \text{im}(1 + T^*). \tag{2.6}$$

If $\bar{y} \in H_*$ is such that $\langle \bar{y}, y \rangle \neq 0$, then $\langle \bar{y}^2, \phi(y) \rangle = \langle \bar{y} \otimes \bar{y}, y \otimes y + \text{im}(1 + T^*) \rangle \neq 0$. Thus $\bar{y}^2 \neq 0$.

If y is indecomposable, then there is an element $\bar{y}$ in PH_* such that $\langle \bar{y}, y \rangle \neq 0$, so that $0 \neq \bar{y}^2 \in PH_*$, implying the existence of a non-primitive generator of H^* whose degree is

$4k$, which is impossible. Thus y must be decomposable. Since x is primitive, so is y. By

the exact sequence

$$0 \to P(\xi H^*) \to PH^* \to QH^* \tag{2.7}$$

$y \in P(\xi H^*)$. Suppose now that y projects to a non-zero element of $Q(\xi H^*)$. Since y

projects to a non-zero element of $Q(\xi H^*)$, there is an indecomposable element y_1 such

that $y = y_1^2$; for if y_1 were decomposable, this would imply that y is decomposable in ξH^*.

Choose a Borel decomposition for H^*. Then $y_1 = z_1 + \cdots + z_m + d$, where the z_i are

certain Borel generators and d is decomposable. Also, since y_1^2 is indecomposable in ξH^*,

at least one of the z_i has square non-zero; assume $z_1^2 \neq 0$. Then as algebras

$$H^* \cong \frac{Z_2[z_1]}{(\varepsilon z_1^{2^r})} \otimes A, \tag{2.8}$$

where $r \geq 2$, $\varepsilon \in Z_2$ and A is an algebra. If $\varepsilon = 0$, this means z_1 has infinite height; if

$\varepsilon = 1$, then z_1 has height 2^r. Dually, as coalgebras

$$H_* \cong \frac{\Gamma[\bar{z}_1]}{(\varepsilon \gamma_{2^r}(\bar{z}_1))} \otimes A^*, \tag{2.9}$$

where $\langle \bar{z}_1, z_i \rangle = \delta_{1i}$ (δ_{ij} is the Kronecker delta). Thus there is an element $\bar{z}_1 \in PH_*$ and

an element $\gamma_2(\bar{z}_1)$ such that $\overline{\Delta}\gamma_2(\bar{z}_1) = \bar{z}_1 \otimes \bar{z}_1$ and $\langle \gamma_2(\bar{z}_1), y \rangle \neq 0$. The previous discussion

implies $\gamma_2(\bar{z}_1)^2 \neq 0$. We have $\overline{\Delta}(\gamma_2(\bar{z}_1)^2) = \bar{z}_1^2 \otimes \bar{z}_1^2$. If $\bar{z}_1^2 = 0$, this implies that $\gamma_2(\bar{z}_1)^2$ is

primitive, and hence dual to an indecomposable cohomology element of degree $4k$. This is impossible since there are no non-primitive generators of H^* in degree $4k > n$. Hence

$\bar{z}_1^2 \neq 0$; but $\bar{z}_1^2 \in PH_*$ since $\bar{z}_1 \in PH_*$. This implies the existence of a non-primitive

generator of H^* of degree $2k$. This is also impossible since $2k > n$. The only possible

conclusion is that y projects to zero in $Q(\xi H^*)$. By the exact sequence

$$0 \to P(\xi^2 H^*) \to P(\xi H^*) \to Q(\xi H^*) \tag{2.10}$$

$y \in P(\xi^2 H^*)$. In particular, $y \in \xi^2 H^*$, which was assumed not to be true. This completes

the proof of CASE 1.

CASE 2: $y^2 \neq 0$. By the Adem relations the following is obtained

$$0 \neq y^2 = Sq^{2k} Sq^1 x = Sq^2 Sq^1 Sq^{2k-2} x. \tag{2.11}$$

Therefore $Sq^{2k-2} x$ is a non-zero primitive element of degree $2(2k-1)-1$ such that

$Sq^1(Sq^{2k-2} x) \neq 0$. Furthermore, the degree of $Sq^1(Sq^{2k-2} x)$ is $4k-2$, so it must be

the case that $Sq^1(Sq^{2k-2} x) \notin \xi^2 H^*$. But $2k-1 > k$ and $2k-1$ is not a power of two; this

is a contradiction to the choice of k. QED

The study of QH^* in even degrees begins with the next lemma. For convenience, the

following definition is useful.

DEFINITION 2.3. *For positive integers r and k, $d(r,k) = 2^r(2k+1)$.*

Note that any positive, even integer can be uniquely expressed as either 2^r or $d(r,k)$ for

some positive integers r and k.

The Steenrod algebra $\mathcal{A}(2)$ acts on the cohomology of X. Suppose $x \in H^*$, $t \in H_*$

and $\alpha \in \mathcal{A}(2)$. Then there is an adjoint action of the Steenrod algebra on H_* and $t\alpha$ is

defined as the element having the property that for each $x \in H^*$, $\langle t\alpha, x \rangle = \langle t, \alpha x \rangle$.

LEMMA 2.4.   If n is as in Lemma 2.2 and if for some $r, k \geq 1$, $2^{r+1}k > n$, then

$QH^{d(r,k)} = 0$.

PROOF: The proof will be by upward induction on r. If $r > 1$, assume $QH^{d(l,k)} = 0$ for

all l and k satisfying $1 \leq l < r$, $k \geq 1$ and $2^{l+1}k > n$. If $r = 1$, there are no inductive

assumptions. For a given r, assume the lemma is false and pick the largest k such that

$2^{r+1}k > n$ and $QH^{d(r,k)} \neq 0$. Pick a primitive element x with non-zero projection in

$QH^{d(r,k)}$; x may be chosen to be primitive since $d(r,k) > 2^{r+1}k > n$. The goal of the

proof is to obtain a contradiction by applying a secondary operation associated to the

factorization $Sq^{d(r,k)+1} = Sq^1 Sq^{d(r,k)}$. It would be convenient if $Sq^{d(r,k)}x = 0$; but this

is not necessarily the case. Instead it will be shown that $Sq^{d(r,k)}x \in \overline{\xi H^*} \cdot \overline{\xi H^*}$, and the

operation used will be the ladder Toda bracket operation that appears in Lin [9]. This

operation will produce a non-primitive generator of degree $d(r+1,k)$, which cannot happen

since $d(r+1,k) > n$.

There is a factorization of the Steenrod algebra given by

$$Sq^{d(r,k)} = Sq^{2^r} Sq^{2^{r+1}k} + \sum_{i=0}^{r-1} Sq^{2^i} \alpha_i \tag{2.12}$$

for certain $\alpha_i \in \mathcal{A}(2)$. It will be shown that $x^2 = Sq^{d(r,k)}x \in \overline{\xi H^*} \cdot \overline{\xi H^*}$ by showing that

the right hand side above is contained in $\overline{\xi H^*} \cdot \overline{\xi H^*}$.

The degree of $Sq^{2^{r+1}k}x$ is $2^r(2(2k)+1) = d(r,2k)$. By the choice of k, $Sq^{2^{r+1}k}x$ is

decomposable. Then since $Sq^{2^{r+1}k}x$ is primitive, $Sq^{2^{r+1}k}x = y^2$ for some $y \in H^*$ of degree

$2^{r-1}(2(2k)+1) = d(r-1,2k)$. If $r = 1$, the degree of y is odd, and y may be written as

a sum $w + d$, where $w \in PH^*$ and $d \in DH^*$. If $r > 1$, then the degree of y is $d(r-1,2k)$

and $2^{(r-1)+1}(2k) = 2^{r+1}k > n$, so by the inductive hypothesis, $y = d$ for some $d \in DH^*$.

In the case that $r = 1$, $w^2 = Sq^{4k+1}w = Sq^1 Sq^{4k}w$ and the degree of $Sq^{4k}w$ is $8k+1$.

Since $8k+1 > 4k > n$, Lemma 2.2 implies that $w^2 = Sq^1(Sq^{4k})w \in \xi^2 H^* \subset \overline{\xi H^*} \cdot \overline{\xi H^*}$. If

$d = \sum_j a_j b_j$, then $d^2 = \sum_j a_j^2 b_j^2$, so that $d^2 \in \overline{\xi H^*} \cdot \overline{\xi H^*}$. This implies in either case that

$Sq^{2^{r+1}k}x = y^2 \in \overline{\xi H^*} \cdot \overline{\xi H^*}$, and thus by the Cartan formula that $Sq^{2^r} Sq^{2^{r+1}k}x \in \overline{\xi H^*} \cdot \overline{\xi H^*}$.

The degree of $\alpha_0 x$ is $2^{r+1}(2k+1) - 1$, which by hypothesis is greater than n. Because

$2^{r+1}(2k+1)$ is not a power of two, Lemma 2.2 implies that $Sq^1 \alpha_0 x \in \xi^2 H^* \subset \overline{\xi H^*} \cdot \overline{\xi H^*}$.

For $1 \leq i \leq r-1$, the degree of $\alpha_i x$ is $2^r(2k+1) + 2^r(2k+1) - 2^i = 2^i(2(2^{r+1-i}k +$

$\sum_{j=0}^{r-1-i} 2^j) + 1) = d(i, 2^{r+1-i}k + \sum_{j=0}^{r-1-i} 2^j)$. Also, $\alpha_i x$ is primitive. Since $2^{i+1}(2^{r+1-i}k +$

$\sum_{j=0}^{r-1-i} 2^j) = 2^{r+2}k + \sum_{j=i+1}^{r} 2^j > n$, the inductive hypothesis implies that $\alpha_i x = y_i^2$ for some element $y_i \in H^*$ of degree $2^{i-1}(2(2^{r+1-i}k + \sum_{j=0}^{r-1-i} 2^j) + 1) = d(i-1, 2^{r+1-i}k +$

$\sum_{j=0}^{r-1-i} 2^j)$. If $i = 1$, then $Sq^2\alpha_1 x = Sq^2 y_1^2 = (Sq^1 y_1)^2$. In this case the degree of y_1 is

$2^r(2k+1) - 1$; y_1 may be written as a sum $w_1 + d_1$, where $w_1 \in PH^*$ and $d_1 \in DH^*$. Then

$Sq^1 y_1 = Sq^1 w_1 + Sq^1 d_1$, which is decomposable by Lemma 2.2 and the Cartan formula.

Thus $Sq^2\alpha_1 x = (Sq^1 y_1)^2 \in \overline{\xi H^*} \cdot \overline{\xi H^*}$. If $2 \leq i \leq r - 1$, then since the degree of y_i is

$d(i-1, 2^{r+1-i}k + \sum_{j=0}^{r-1-i} 2^j)$ and since $2^{(i-1)+1}(2^{r+1-i}k + \sum_{j=0}^{r-1-i} 2^j) = 2^{r+1}k + \sum_{j=i}^{r-1} 2^j >$

n, the inductive hypothesis implies that y_i is decomposable. Thus $\alpha_i x = y_i^2 \in \overline{\xi H^*} \cdot \overline{\xi H^*}$

and $Sq^{2^i}\alpha_i x \in \overline{\xi H^*} \cdot \overline{\xi H^*}$.

It has been shown that $Sq^{d(r,k)} x \in \overline{\xi H^*} \cdot \overline{\xi H^*}$. Furthermore, x is primitive. Under these

circumstances Lin [9] has constructed a secondary operation ϕ defined on x associated to

the factorization $Sq^{d(r,k)+1} = Sq^1 Sq^{d(r,k)}$. This operation has the property that

$$\overline{\Delta}\phi(x) = x \otimes x + \text{im } Sq^1 + H^* \otimes \xi H^* + \xi H^* \otimes H^*. \qquad (2.13)$$

Pick $t \in PH_*$ such that $\langle t, x \rangle \neq 0$. Then $tSq^1 = 0$ by Lemma 2.2 and $\langle t, \xi H^* \rangle = 0$ since t

is primitive, so that

$$\langle t^2, \phi(x) \rangle = \langle t \otimes t, \overline{\Delta}\phi(x) \rangle$$

$$= \langle t \otimes t, x \otimes x \rangle \neq 0. \qquad (2.14)$$

This implies the existence of a non-primitive generator of H^* of degree $2^{r+1}(2k + 1) =$

$d(r + 1, k)$. This is impossible since $d(r + 1, k) > n$. QED

LEMMA 2.5.   *For $r \geq 1$, let $2^{r+1}l$ be the highest degree such that l is not a power of two*

in which there is a non-primitive generator of degree congruent to $0 \bmod 2^{r+1}$. Let $d(r,k)$

be the highest degree in which there is a non-primitive generator of degree congruent to

$2^r \bmod 2^{r+1}$. Then $2k \leq l$; in particular, every indecomposable element of H^* in a degree

not equal to a power of two has a primitive representative.

PROOF: The statement of Lemma 2.5 may seem a bit mysterious at first reading. But the

statement means essentially that the highest degree not equal to a power of two in which

there is a non-primitive generator, if it exists at all, becomes unbounded. For example,

suppose the highest degree congruent to $0 \bmod 4$ in which there is a non-primitive generator

is $4l$ and the highest degree congruent to $2 \bmod 4$ in which there is a non-primitive generator

is $d(1,k) = 4k + 2$. Then $2k \leq l$, which means $4k \leq 2l$, and hence $4k + 2 < 4l$. The upshot

is that the highest degree not equal to a power of two in which there is a non-primitive

generator is in a degree congruent to $0 \bmod 4$. This covers the case where $r = 1$. Arguing

similarly for each successive value of r, it can be shown that the highest degree not equal

to a power of two in which there is a non-primitive generator is in a degree congruent to

$0 \bmod 2^{r+1}$ for any $r \geq 1$. This implies that no such generator exists.

The proof is similar to the proof of Lemma 2.4; it follows by upward induction on r.

Assume by induction that the proposition holds for all values up to $r - 1$, except in the case that $r = 1$, where no inductive assumption is made. Note that this implies that the

highest degree not equal to a power of two in which there is a non-primitive generator is

congruent to 0 mod 2^r, and hence this degree is equal to either $2^{r+1}l$ or $d(r,k)$. Assume

that $2k > l$. It will be shown that this is not possible. Pick $\hat{x} \in QH^{d(r,k)}$ which corresponds

to a non-primitive generator. As in the proof of Lemma 2.4, it is desirable to apply the

ladder Toda bracket operation associated to the factorization $Sq^{d(r,k)+1} = Sq^1 Sq^{d(r,k)}$. In

order to do so, it is necessary to find an $\mathcal{A}(2)$ sub-Hopf algebra B of H^* such that for some

x which projects to $\hat{x}$, $Sq^{d(r,k)}x \in \overline{B} \cdot \overline{B}$ and $\overline{\Delta}(x) \in B \otimes B$.

Let B^n denote the $\mathcal{A}(2)$ sub-algebra of H^* generated by the set S^n of all primitive

elements and all elements of degree less than or equal to n. Since S^n is a coalgebra, B^n is

a Hopf algebra, and H^* is filtered by an increasing sequence of $\mathcal{A}(2)$ Hopf algebras

$$B^0 \subset B^1 \subset \cdots \subset B^n \subset B^{n+1} \subset \cdots \subset H^*. \qquad (2.15)$$

Observe that $B^{n+1} /\!/ B^n$ is a primitively generated Hopf algebra, and is generated over

$\mathcal{A}(2)$ by the elements of degree $n + 1$. Let τ_n be the composite

$$\overline{B^n} \to \overline{H^*} \to QH^*, \qquad (2.16)$$

and let $\operatorname{im} \tau_n = F_n Q$. Then QH^* is filtered by $\mathcal{A}(2)$ modules

$$F_0 Q \subset F_1 Q \subset \cdots \subset F_n Q \subset F_{n+1} Q \subset \cdots \subset QH^*. \qquad (2.17)$$

If $\hat{y} \in F_0 Q$ then $\hat{y}$ has a primitive representative. Therefore $\hat{x} \notin F_0 Q$. Suppose $\hat{x} \in F_{n+1} Q$

and $\hat{x} \notin F_n Q$. Pick $x' \in B^{n+1}$ such that $\tau_{n+1}(x') = \hat{x}$. Then

$$x' = (\sum_{i=1}^{m} \alpha_i x_i) + q, \qquad (2.18)$$

where $\alpha_i \in \mathcal{A}(2)$, x_i is either primitive or has degree less than or equal to $n + 1$, and

$q \in \overline{B^{n+1}} \cdot \overline{B^{n+1}}$. Then $\overline{\Delta}(x_i) \in B^n \otimes B^n$ so that $\overline{\Delta}(\alpha_i x_i) \in B^n \otimes B^n$. Set $x = \sum_{i=1}^{m} \alpha_i x_i$

and let $B = B^n + \xi H^*$; then $\overline{\Delta}(x) \in B \otimes B$. Note that since $\hat{x}$ is non-zero in $Q(\lambda H^*)$ and

$\hat{x} \notin F_n Q$, there is an element $t^2 \in PH_*$ such that $\langle t^2, x \rangle \neq 0$ and $\langle t^2, B \rangle = 0$.

Using the factorization

$$Sq^{d(r,k)} = Sq^{2^r} Sq^{2^{r+1}k} + \sum_{i=0}^{r-1} Sq^{2^i} \alpha_i \qquad (2.19)$$

it will be shown that $Sq^{d(r,k)} x \in \overline{B} \cdot \overline{B}$ by showing that the right hand side above is

contained in $\overline{B} \cdot \overline{B}$.

The degree of $Sq^{2^{r+1}k} x$ is $2^r(2(2k) + 1) = d(r, 2k)$. Since by assumption $2^{r+1}(2k) >$

$2^{r+1} l$, and since $2^{r+1}(2k) > d(r, k)$, Lemma 2.4 implies that $Sq^{2^{r+1}k} x$ must be decompos-

able. Let $F(s)$ denote the sub-Hopf algebra of B generated by elements of degree less than

s, and let $A = F(d(r, 2k))$. Then since $\overline{\Delta}(Sq^{2^{r+1}k} x) \in A \otimes A$, $\{Sq^{2^{r+1}k} x\}$ is decomposable

and primitive in $H^*/\!/A$. This implies that $\{Sq^{2^{r+1}k} x\} = \{z\}^2$ in $H^*/\!/A$. Let $z \in H^*$ be

a representative of $\{z\}$. The degree of z is $2^{r-1}(2(2k) + 1) = d(r - 1, 2k)$. The highest degree in which there is a non-primitive generator in a degree congruent to zero mod 2^r is

either $2^r(2k+1)$ or $2^r(2l)$. If $r > 1$, then by the inductive hypothesis, since $2(2k) > 2k+1$

and since $2(2k) > 2l$, the element z can be written as a sum $w + d$, where $w \in PH^*$

and $d \in DH^*$. If $r = 1$, then the degree of z is odd and w and d may be chosen simi-

larly. Then $\{z\}^2 = \{w^2\} + \{d^2\}$ and $d^2 \in F(d(r, 2k))$, so in fact, $\{Sq^{2^{r+1}k}x + w^2\} = 0$

in $H^* /\!/ A$. Then $Sq^{2^{r+1}k}x + w^2 \in \overline{A} \cdot H^*$. Also, $\overline{\Delta}(w^2) = 0$ since w is primitive. Thus

$\overline{\Delta}(Sq^{2^{r+1}k}x + w^2) \in A \otimes A$. Following Milnor and Moore [**9**], there is an isomorphism of

left A modules given by the composition

$$\theta : H^* \xrightarrow{\Delta} H^* \otimes H^* \xrightarrow{\eta \otimes \pi} A \otimes H^* /\!/ A, \tag{2.20}$$

where η is a splitting of the exact sequence $0 \to A \to H^*$ and π is the canonical projection.

By applying the θ to $Sq^{2^{r+1}k}x + w^2$, the result is $\theta(Sq^{2^{r+1}k}x + w^2) = (Sq^{2^{r+1}k}x + w^2) \otimes 1$.

This implies that $Sq^{2^{r+1}k}x + w^2 \in A$. But since $A \subset B$, and since A is generated by

elements of B whose degree is less than the degree of $Sq^{2^{r+1}k}x + w^2$, it must be the case

that $Sq^{2^{r+1}k}x + w^2$ is decomposable in B, i.e. $Sq^{2^{r+1}k}x + w^2 \in \overline{B} \cdot \overline{B}$. Since w is primitive,

$w \in B^n \subset B$. This implies that $w^2 \in \overline{B} \cdot \overline{B}$. Therefore, $Sq^{2^{r+1}k}x \in \overline{B} \cdot \overline{B}$ and hence

$Sq^{2^r}Sq^{2^{r+1}k}x \in \overline{B} \cdot \overline{B}$ by the Cartan formula.

The degree of $\alpha_0 x$ is $2^{r+1}(2k+1)-1 = d(r+1, k)-1$. Since the degree is odd, $\alpha_0 x$ may

be written as a sum $\alpha_0 x = y + a$, where $y \in PH^*$ and $a \in DH^*$. Let $C = F(d(r+1, k)-1)$.

Then $\{a\}$ is primitive and decomposable in $H^*/\!/C$. Since the degree of $\{a\}$ is odd, $\{a\} = 0$.

Thus $\{\alpha_0 x + y\} = 0$ in $H^*/\!/C$, and $\overline{\Delta}(\alpha_0 x + y) \in C \otimes C$. As in the previous paragraph, it may

be concluded that $\alpha_0 x + y \in \overline{B} \cdot \overline{B}$. The degree of y is $d(r+1, k) - 1$ and $d(r+1, k) - 1 > 2^{r+1}l$

and $d(r+1, k) - 1 > d(r, k)$. Lemma 2.2 implies that $Sq^1 y \in \xi^2 H^* \subset \overline{B} \cdot \overline{B}$, so that

$Sq^1 \alpha_0 x \in \overline{B} \cdot \overline{B}$.

For $1 \le i \le r - 1$, the degree of $\alpha_i x$ is $2^i(2(2^{r+1-i}k + \sum_{j=0}^{r-1-i} 2^j) + 1) = d(i, 2^{r+1-i}k +$

$\sum_{j=0}^{r-1-i} 2^j)$. Set $p_i = 2^{r+1-i}k + \sum_{j=0}^{r-1-i} 2^j$. Since $2^{i+1}(2^{r+1-i}k + \sum_{j=0}^{r-1-i} 2^j) > 2^{r+1}(2k) >$

$2^{r+1}l$ and $2^{i+1}(2^{r+1-i}k + \sum_{j=0}^{r-1-i} 2^j) > d(r, k)$, Lemma 2.4 implies that $\alpha_i x$ is decompos-

able. Let $A_i = F(d(i, p_i))$. Then $\{\alpha_i x\}$ is primitive and decomposable in $H^*/\!/A_i$. Thus

$\{\alpha_i x\} = \{z_i\}^2$ for some $z_i \in H^*$ of degree $d(i-1, p_i) = 2^{i-1}(2(2^{r+1-i}k + \sum_{j=0}^{r-1-i} 2^j) + 1)$.

The highest degree in which there is a non-primitive generator in a degree congruent

to zero mod 2^i is either $2^i(2^{r-i}(2k+1))$ or $2^i(2^{r+1-i}l)$. If $i > 1$, then since $2p_i =$

$2(2^{r+1-i}k + \sum_{j=0}^{r-1-i} 2^j) > 2^{r-i}(4k) > 2^{r-i}(2k+1)$ and $2p_i = 2(2^{r+1-i}k + \sum_{j=0}^{r-1-i} 2^j) >$

$2^{r+1-i}(2k) > 2^{r+1-i}l$, the inductive hypothesis implies that z_i may be written as a sum

$w_i + d_i$, where $w_i \in PH^*$ and $d_i \in DH^*$. If $i = 1$, then the degree of z_1 is odd, and so

$z_1 = w_1 + d_1$, with $w_1 \in PH^*$ and $d_1 \in DH^*$. In all cases $\{z_i\}^2 = \{w_i^2\} + \{d_i^2\}$, and

$d_i^2 \in F(p_i) \subset A_i$ so that $\{\alpha_i x + w_i^2\} = 0$. Since w_i^2 is primitive, $\overline{\Delta}(\alpha_i x + w_i^2) \in A_i \otimes A_i$

and, as before, the Milnor-Moore isomorphism may be applied to the conclusion that

$\alpha_i x + w_i^2 \in \overline{B} \cdot \overline{B}$. Since w_i is primitive, $w_i^2 \in \overline{B} \cdot \overline{B}$. Therefore $\alpha_i x \in \overline{B} \cdot \overline{B}$, and hence

$Sq^{2^r} \alpha_i \in \overline{B} \cdot \overline{B}$ by the Cartan formula.

It has been shown that $Sq^{d(r,k)} x \in \overline{B} \cdot \overline{B}$. Furthermore, B was constructed with the

property that $\overline{\Delta} x \in B \otimes B$. Under these circumstances Lin [**9**] has constructed a secondary

operation ϕ defined on x associated to the factorization $Sq^{d(r,k)+1} = Sq^1 Sq^{d(r,k)}$. This

operation has the property that

$$\overline{\Delta}\phi(x) = x \otimes x + \text{im}\, Sq^1 + H^* \otimes B + B \otimes H^*. \tag{2.21}$$

Choose $t^2 \in PH_*$ such that $\langle t^2, x \rangle \neq 0$ and $\langle t^2, B \rangle = 0$. Then $(t^2)Sq^1 = 0$ by the Cartan

formula. Hence

$$\langle t^4, \phi(x) \rangle = \langle t^2 \otimes t^2, \overline{\Delta}\phi(x) \rangle = \langle t^2 \otimes t^2, x \otimes x \rangle \neq 0. \tag{2.22}$$

Therefore $t^4 \neq 0$. But $t^4 \in PH_*$, so it is dual to a non-primitive generator of degree

$d(r+1,k)$. This is a contradiction to the choice of l because $d(r+1,k)$ is congruent to

zero mod 2^{r+1} and it was assumed that $2k > l$, which implies $d(r+1,k) > 2^{r+1}l$. QED

It is now easy to prove the following proposition, which is the main result of this

section.

PROPOSITION 2.6. $QH^{d(r,k)} = 0$ for all r and k greater than zero.

PROOF: Lemma 2.5 implies that there are no non-primitive generators. Under these

conditions, the Proposition follows from Lemma 2.4. QED

§3 Initial study of QH^{odd}

In §2 it was shown that the only even degree indecomposables are in degrees which are powers of two. In this section the study of odd degree indecomposables begins. The main results of this section are Propositions 3.3, 3.5, 3.6 and 3.9. The combination of these propositions summarized gives the result that $QH^{d(r,k)-1} = 0$ for all $r \geq 2$ and $k \geq 1$, and for all $r \geq 1$ when k is not a power of two. This implies that the only odd degree indecomposables must lie in degrees $d(1, 2^s) - 1$ or $2^s - 1$, which are powers of two plus or minus one. A summary appears in the table below, where the right hand column indicates the degrees in which $QH^* = 0$ according to the corresponding proposition in the left hand column.

Proposition 3.3	$d(r, k) - 1$ for $r \geq 1$, $k \geq 1$ and not a power of two
Proposition 3.5	$d(r, 2^s) - 1$ for $r \geq 3$ and $s \geq 0$
Proposition 3.6	$d(2, 2^s) - 1$ for $s \geq 1$
Proposition 3.9	$d(2, 2^0) - 1$

The other results of this section, Lemmas 3.1, 3.2, 3.4, 3.7, 3.8 and 3.10, give some control over the action of $\mathcal{A}(2)$ on H^* and on $H^*(\Omega X)$. This control is essential in the application of the secondary operation arguments that appear in the four propositions mentioned above.

LEMMA 3.1.   *If* $x \in H^{odd}$, *then* $x^2 = 0$.

PROOF: It will first be shown that the squaring map is zero when restricted to odd dimensional primitives.

Let $y \in PH^{2k-1}$. Then $y^2 = Sq^{2k-1}y = Sq^1 Sq^{2k-2}y$ and $Sq^{2k-2}y$ has degree $4k - 3 = 2(2k-1) - 1$. Lemma 2.2 combined with Proposition 2.5 implies that $y^2 = Sq^1(Sq^{2k-2}y) \in \xi^2 H^*$. But $\deg y^2 = 4k - 2$, so it must be the case that $y^2 = 0$.

Now $\overline{H}^{odd} \subset \overline{PH}^{odd} \cdot H^*$. Let $x \in H^{odd}$; we can write $x = \sum_i y_i z_i$, with $y_i \in PH^{odd}$ and $z_i \in H^*$. Then $x^2 = \sum_i y_i^2 z_i^2 = 0$. QED

LEMMA 3.2.   *If* k *is not a power of two, then* $Sq^1 PH^{2k-1} = 0$.

PROOF: Suppose that $x \in PH^{2k-1}$ and $Sq^1 x = y$. Then Lemma 2.2 implies that $y \in \xi^2 H^*$; in particular, y is decomposable and primitive. Therefore, by the exact sequence

$$0 \to P(\xi H^*) \to PH^* \to QH^*, \tag{3.1}$$

$y \in P(\xi H^*)$. If $y = z^2$, and z is decomposable, then y must be decomposable in ξH^*; so by the exact sequence

$$0 \to P(\xi^2 H^*) \to P(\xi H^*) \to Q(\xi H^*), \tag{3.2}$$

$y \in P(\xi^2 H^*)$. This argument may be repeated as often as necessary to the conclusion that

$y = w^{2^i}$ for some indecomposable element w and some $i \geq 1$. Since w is indecomposable

and $\deg w = \frac{k}{2^{i-1}}$ is not a power of two, it must be the case that $\deg w$ is odd, and by

Proposition 3.1, $w^2 = 0$. Thus $y = 0$. QED

In Proposition 3.3 below, a secondary operation is constructed with the intent of

showing the existence of a certain non-primitive generator. But this generator occurs in

an even degree which is not a power of two, which is impossible. This will give a desired

contradiction. In order to show the existence of such a generator, the indeterminacy of the

operation must be carefully analyzed. This analysis makes up the bulk of the proof.

PROPOSITION 3.3.   For all $r \geq 1$ and all $k \geq 1$ such that k is not a power of two,

$QH^{d(r,k)-1} = 0$.

PROOF: Suppose the proposition is false. Pick the smallest r such that the proposition

fails; and pick the largest $k \neq 2^s$ such that $QH^{d(r,k)-1} \neq 0$. Pick an indecomposable

element $x \in PH^{d(r,k)-1}$; such an element exists since all odd degree indecomposables have primitive representatives. There is a factorization of the Steenrod algebra given by

$$Sq^{d(r,k)} = Sq^{2^r} Sq^{2^{r+1}k} + \sum_{i=0}^{r-1} Sq^{2^i} \alpha_i. \tag{3.3}$$

Consider the two stage Postnikov system

$$K(Z_2, n_0 - 1, \ldots, n_r - 1)$$

$$\downarrow j$$

$$\begin{array}{ccc} & E & \xrightarrow{v} & K(Z_2, d(r+1,k) - 2) \qquad (3.4) \\ & \downarrow p & \\ X \xrightarrow{f} & K(Z_2, d(r,k) - 1) & \xrightarrow{g} & K(Z_2, n_0, \ldots, n_r) \end{array}$$

where f represents x, $g^* = \begin{pmatrix} \alpha_0 & \cdots & \alpha_{r-1} & Sq^{2^{r+1}k} \end{pmatrix}$ and $n_i = d(r+1,k) - 2^i - 1$. The

map v comes from the relation (3.3) above and corresponds to a secondary operation ϕ.

The class v satisfies $\overline{\Delta}(v) = p^* \iota_{d(r,k)-1} \otimes p^* \iota_{d(r,k)-1}$ and $j^*(v) = \sum_{i=0}^r Sq^{2^i} \iota_{n_i-1}$.

In order to show that the lifting $\tilde{f}$ exists, we need to show that $Sq^{2^{r+1}k}x = 0$ and

$\alpha_i x = 0$ for each i. The degree of $Sq^{2^{r+1}k}x$ is $2^r(2(2k)+1)-1 = d(r,2k)-1$; by the choice of

k, $Sq^{2^{r+1}k}x$ is decomposable. As $Sq^{2^{r+1}k}x$ is primitive, it must be a square. But the degree

of $Sq^{2^{r+1}k}x$ is odd, so $Sq^{2^{r+1}k}x = 0$. The degree of $\alpha_0 x$ is $2^{r+1}(2k+1) - 2$; since there

are no indecomposables in that degree, $\alpha_0 x$ must be decomposable. It is also primitive,

so $\alpha_0 x$ is the square of an odd degree class. Therefore, $\alpha_0 x = 0$. For $1 \le i \le r - 1$, the degree of $\alpha_i x$ is $2^i(2(2^{r+1-i}k + \sum_{j=0}^{r-1-i} 2^j) + 1) - 1 = d(i, 2^{r+1-i}k + \sum_{j=0}^{r-1-i} 2^j) - 1$. The

choice of r implies that $\alpha_i x$ is decomposable. But $\alpha_i x$ is also primitive and of odd degree,

so $\alpha_i x = 0$. Therefore, the lifting $\tilde{f}$ exists.

Because x is primitive, f is an H-map. Since D_f, the H-deviation of f, is homotopic

to $pD_{\tilde{f}}$, $D_{\tilde{f}}$ factors through the fibre

$$X \wedge X \xrightarrow{D} K(Z_2, n_0 - 1, \ldots, n_r - 1) \xrightarrow{j} E. \qquad (3.5)$$

The matrix $[D]$ satisfies the relation $(\Omega g)h_1(f) = (1 + T^*)[D]$, where $h_1(f)$ is the h_1-

deviation of f and Ωg is given by the matrix

$$\Omega g = \begin{pmatrix} \alpha_0 \\ \cdot \\ \cdot \\ \cdot \\ \alpha_{r-1} \\ Sq^{2^{r+1}k} \end{pmatrix}. \qquad (3.6)$$

The secondary operation ϕ associated to v satisfies the formula

$$\overline{\Delta}\phi(x) = x \otimes x + D_{\tilde{f}}^*(v) = x \otimes x + \sum_{i=0}^{r} Sq^{2^i}([D_i]), \qquad (3.7)$$

where the $[D_i]$ are elements of $H^* \otimes H^*$ and are the entries of the column matrix $[D]$. We

wish to show that there is a sub-vector space of $H^* \otimes H^*$ which contains the indeterminacy

$\sum_{i=0}^{r} Sq^{2^i}([D_i])$, but does not contain $x \otimes x$ (modulo decomposables). In other words, the

element $x \otimes x$ is not "cancelled" by the indeterminacy.

Let B be the sub-vector space of $H^* \otimes H^*$ spanned by elements which are either decomposable on one factor, contained in $\mathrm{im}\,(1 + T^*)$, or are of the form $a \otimes b$, where $\deg a \neq \deg b$. Suppose it can be shown for each i that $Sq^{2^i}[D_i] \in B$. Then

$$\overline{\Delta}\phi(x) = x \otimes x + B. \tag{3.8}$$

Choose $t \in PH_{d(r,k)-1}$ such that $\langle t, x \rangle \neq 0$. Then $\langle t \otimes t, B \rangle = 0$ since $\langle t, DH^* \rangle = 0$, and

$$\langle t^2, \phi(x) \rangle = \langle t \otimes t, \overline{\Delta}\phi(x) \rangle = \langle t \otimes t, x \otimes x \rangle \neq 0. \tag{3.9}$$

This implies that $0 \neq t^2 \in PH_{d(r+1,k)-2}$; but since $QH^{d(r+1,k)-2} = QH^{d(1,2^r k + \sum_{j=0}^{r-2} 2^j)} = 0$ by Proposition 2.6, it must be the case that $t^2 = 0$. This would complete the proof by contradiction.

It remains to be shown for each i that $Sq^{2^i}[D_i] \in B$. There are three cases, $i = 0$, $1 \leq i < r$, and $1 \leq i = r$.

CASE 1: $i = 0$. We can write

$$[D_0] \in H^{d(r,k)-2} \otimes H^{d(r,k)-1} + H^{d(r,k)-1} \otimes H^{d(r,k)-2} + \sum_{m,n} H^m \otimes H^n, \tag{3.10}$$

where the m and n satisfy $|m - n| > 1$. Then since

$$d(r,k) - 2 = \begin{cases} 4k & \text{if } r = 1, \\ 2^{r+1}k + \sum_{j=1}^{r-1} 2^j & \text{if } r \geq 2, \end{cases} \tag{3.11}$$

and k is not a power of two, $H^{d(r,k)-2} \subset DH^{d(r,k)-2}$, and it is not difficult to see that $Sq^1[D_0] \in B$. This completes the proof of CASE 1.

CASE 2: $1 \leq i < r$. We can write

$$[D_i] \in \sum_{p=-2^{i-1}}^{2^{i-1}} H^{d(r,k)-1-2^{i-1}+p} \otimes H^{d(r,k)-1-2^{i-1}-p} + \sum_{m,n} H^m \otimes H^n, \qquad (3.12)$$

where the m and n satisfy $|m - n| > 2^i$. For $0 \leq p \leq 2^{i-1}$, $d(r,k) - 1 - 2^{i-1} - p = 2^{r+1}k + (2^r - 2^{i-1} - p) - 1$ and $0 < (2^r - 2^{i-1} - p) < 2^r$. Then if $d(r,k) - 1 - 2^{i-1} - p$ is even, it is not a power of two; and if $d(r,k) - 1 - 2^{i-1} - p$ is odd, it is equal to $d(q,s) - 1$ for some q and s where $q \leq r - 1$. Proposition 2.6 in the even case and the choice of r in the odd case imply that $H^{d(r,k)-1-2^{i-1}-p} \subset DH^{d(r,k)-1-2^{i-1}-p}$. The same argument implies that $H^{d(r,k)-1-2^{i-1}+p} \subset DH^{d(r,k)-1-2^{i-1}+p}$ for $-2^{i-1} \leq p \leq 0$. Now it is evident that $Sq^{2^i}[D_i] \in B$. This completes the proof of CASE 2.

CASE 3: $1 \leq i = r$. The class $[D_r]$ satisfies the formula

$$(1 + T^*)[D_r] = (\Omega g)h_1(f) \in Sq^{2^{r+1}k}H^{d(r,k)-2}(X \wedge X). \qquad (3.13)$$

If $r = 1$, then $Sq^{2^{r+1}k}H^{d(r,k)-2}(X \wedge X) = Sq^{4k}H^{4k}(X \wedge X) = \xi H^{4k}(X \wedge X) \subset B$. If $r \geq 2$, then $d(r,k) - 2 = 2^{r+1}k + \sum_{j=1}^{r-1} 2^j$. Since k is not a power of two, the dyadic expansion of $d(r,k) - 2$ has at least three terms. By Proposition 2.6, the only even degree indecomposables of H^* lie in degrees of a power of two. Therefore any non-zero elements of $H^{d(r,k)-2}(X \wedge X)$ must be non-zero in $QH^{odd} \otimes QH^{odd} + DH^* \otimes H^* + H^* \otimes DH^*$. Considering that $\frac{1}{2}(d(r,k) - 2) = d(r-1,k) - 1$ together with the choice of r, note that a basis element

$y \otimes z$ with non-zero projection in $QH^{odd} \otimes QH^{odd}$ such that $\deg(y \otimes z) = d(r, k) - 2$ has

the property that $\deg y \neq \deg z$. Further analysis will reveal that $|\deg y - \deg z| \geq 2^r$.

This is shown in the following argument.

If the assertion is false, the elements y and z must lie in degrees between $d(r - 1, k) -$

$1 - 2^{r-1} = 2^r k - 1$ and $d(r - 1, k) - 1 + 2^{r-1} = 2^r(k+1) - 1$, where all odd degrees have the

form $(2^r k - 1) + 2c$, for some $1 \leq c < 2^{r-1}$. Suppose that $c = 2^h d$, d odd and $0 \leq h < r - 1$.

Then $(2^r k - 1) + 2c = 2^{h+1}(2(\frac{d-1}{2} + 2^{r-h-2}k) + 1) - 1 = d(h + 1, \frac{d-1}{2} + 2^{n-h-2}k) - 1$. The

fact that $2^h d < 2^{r-1}$ implies that $\frac{d-1}{2} < 2^{r-h-2}$. Therefore, $\frac{d-1}{2} + 2^{r-h-2}k$ cannot be a

power of two and by the choice of r, since $h + 1 < r$, there are no non-zero indecomposables

in these degrees. This proves the assertion that $|\deg y - \deg z| \geq 2^r$.

Assume (without loss of generality) that $\deg y \leq 2^r k - 1$ and $\deg z \geq 2^r(k + 1) - 1$.

Now if $Sq^{2^{r+1}k}(y \otimes z) = \sum_q w_q \otimes w_q'$, then $|\deg w_q - \deg w_q'| \geq 2^r + 2$. This follows

from the Cartan formula and the fact that $|\deg y - \deg z| \geq 2^r$. Then applying Sq^{2^r} to

$\sum_q w_q \otimes w_q'$ cannot yield diagonal terms because of dimensional considerations. (A term is

called diagonal if its two factors have the same degree). Note that any term not contained

in B must be diagonal.

It has been shown that applying Sq^{2^r} to non-zero terms of $(1 + T^*)[D_r]$ cannot yield

diagonal terms for dimensional reasons. Thus if $Sq^{2^r}[D_r]$ contains any diagonal terms, then these terms must come from Sq^{2^r} applied to terms contained in $\ker(1 + T^*)$. Terms

contained in $\ker(1 + T^*)$ have the form $a \otimes b + b \otimes a$ or $c \otimes c$. Applying Sq^{2^r} to such

terms yields terms in $\operatorname{im}(1 + T^*)$ except possibly for the term $Sq^{2^{r-1}} c \otimes Sq^{2^{r-1}} c$. But it

was shown earlier that in this case the choice of r dictates that the degree of c must be

such that $c \in DH^*$ since $d(r, k) - 1 - 2^{r-1} = d(r - 1, 2k) - 1$. Therefore $Sq^{2^r}[D_r] \in B$ as

desired and the proof is complete. QED

The proofs of several of the propositions in this work follow a similar pattern. A sec-

ondary operation is constructed with the desire to find a primitive element in cohomology

with non-zero h_1-deviation. However, in an effort to do this, it is difficult to verify that the

cohomology class obtained is primitive. It is possible to show that the reduced coproduct

is contained in $H^* \otimes DH^* + DH^* \otimes H^*$. This will imply that the class obtained from

the looped operation is not only primitive, but is an h_1-class (can be represented by an

h_1-map). Then we wish to show that the h_2-deviation of this class is non-zero. In order to

analyze the indeterminacy that arises from the construction of the operation, the Cartan

formula of Thomas [16] for secondary operations is used. In order to apply this argument,

the following lemma is needed. Note that if x is a cohomology class, then $h_n(x)$ means

$h_n(f)$, where f is a map representing x. There is some indeterminacy arising from the

choice of h_{n-1} structure map for f, but this indeterminacy doesn't come into play in the next lemma.

LEMMA 3.4.   *Suppose $x \in PH^*$ is an element of degree not congruent to 1 mod 4, and suppose $\sigma^* x$ is indecomposable. Then there is an $\mathcal{A}(2)$ sub-Hopf algebra B_m of $H^*(\Omega X)$ such that the following two conditions hold:*

$$(1) \; h_2(\sigma^* x) \in B_m \otimes B_m$$

$$(2) \; \text{If } QB_m \xrightarrow{i_m} QH^*(\Omega X) \text{ is the map induced}$$

$$\text{by inclusion, then } \widehat{\sigma^* x} \notin \operatorname{im} i_m$$

Note that condition (2) implies the existence of an element $t \in PH^*(\Omega X)$ such that $\langle t, \sigma^* x \rangle \neq 0$ and $\langle t, B_m \rangle = 0$.

PROOF: Let S_m be the set of indecomposable elements of H^* whose degrees are less than or equal to m. Let $\sigma^*(S_m)$ be the subspace of $H^*(\Omega X)$ spanned by the image of the elements of S_m under σ^*. Let B_m be the $\mathcal{A}(2)$ sub-algebra of $H^*(\Omega X)$ generated by the set $\sigma^*(S_m)$. Because $\operatorname{im} \sigma^* \subset PH^*(\Omega X)$, B_m is a coalgebra, and hence a Hopf algebra. There is an increasing (not necessarily exhaustive) filtration of $\mathcal{A}(2)$ Hopf algebras

$$0 = B_0 \subset \cdots \subset B_m \subset B_{m+1} \subset \cdots \subset H^*(\Omega X). \qquad (3.14)$$

Let $i_m : QB_m \to QH^*(\Omega X)$ be induced by inclusion of B_m in $H^*(\Omega X)$. For some m,

$\widehat{\sigma^* x} \notin \operatorname{im} i_m$ and $\widehat{\sigma^* x} \in \operatorname{im} i_{m+1}$. Then there is a decomposition

$$\sigma^* x = \left(\sum_j \alpha_j \sigma^* x_j \right) + d \tag{3.15}$$

where $\alpha_j \in \mathcal{A}(2)$, $x_j \in S_{m+1}$ and $d \in DH^*(\Omega X)$.

By the decomposition above, $d \in \operatorname{im} \sigma^*$. In particular, d is primitive, so by the exact

sequence

$$0 \to P(\xi H^*(\Omega X)) \to PH^*(\Omega X) \to QH^*(\Omega X), \tag{3.16}$$

$d = a^2$ for some class $a^2 \in P(\xi H^*(\Omega X))$. Since $\deg x$ is either 0, 2 or 3 mod 4, the degree of

d is either 1, 2 or 3 mod 4. In the case that $\deg d$ is congruent to 1 or 3 mod 4, $d = a^2 = 0$.

In the case that $\deg d$ is congruent to 2 mod 4, $\deg a$ is odd. If a is decomposable then a^2

is decomposable in $\xi H^*(\Omega X)$, so by the exact sequence

$$0 \to P(\xi^2 H^*(\Omega X)) \to P(\xi H^*(\Omega X)) \to Q(\xi H^*(\Omega X)), \tag{3.17}$$

$d = a^2 \in P(\xi^2 H^*(\Omega X))$. But this is impossible if $\deg d$ is congruent to 2 mod 4. Hence a

is indecomposable and of odd degree. In this case a may be chosen to lie in $\operatorname{im} \sigma^*$. Suppose

$a = \sigma^* b$; then $d = a^2 = Sq^n a = Sq^n \sigma^* b$, where $\deg b = n + 1$ and b is indecomposable.

Now $\sigma^* x$ has the form

$$\sigma^* x = \left(\sum_j \alpha_j \sigma^* x_j \right) + Sq^n \sigma^* b. \tag{3.18}$$

If the degrees of the x_j and of b are odd, then they may be chosen to be primitive since σ^* kills decomposables.

Suppose for some j, $\deg x_j$ is even; then $\deg x_j = 2^s$ for some $s \geq 2$. Suppose for some $r \geq 0$ that $Sq^r x_j$ is indecomposable. If $r > 0$ and is even, then this is impossible since $\deg Sq^r x_j$ is even and the only even degree indecomposables are in degrees of a power of 2. If r is odd, then $Sq^r x_j = Sq^1 Sq^{r-1} x_j$, and $Sq^{r-1} x_j$ is decomposable unless $r = 1$. The only possiblities that remain are that $r = 0$ or $r = 1$. Since $\alpha_j \sigma^* x_j = \sigma^* (\alpha_j x_j)$ and σ^* kills decomposables, it may be concluded that if $\deg x_j$ is even, then either $\alpha_j = Sq^0$ or $\alpha_j = \beta_j Sq^1$ for some $\beta_j \in \mathcal{A}(2)$. Because $\widehat{\sigma^* x} \notin \text{im} \, i_m$, the degree of x must be greater than or equal to $m + 1$. Therefore, if $\alpha_j = Sq^0$ for some j, then $\deg x_j = \deg x = m + 1$. In this case,

$$\sigma^* x = Sq^0 \sigma^* x_j; \qquad x_j = x, \tag{3.19}$$

and $x_j \in PH^*$ as desired. In the case that $\alpha_j \neq Sq^0$ for all j,

$$\sigma^* x = \left(\sum_j \gamma_j \sigma^* y_j \right) + Sq^n \sigma^* b, \tag{3.20}$$

where

$$(\gamma_j, y_j) = \begin{cases} (\alpha_j, x_j) & \text{if } \deg x_j \text{ is odd;} \\ (\beta_j, Sq^1 x_j) & \text{if } \deg x_j \text{ is even.} \end{cases} \tag{3.21}$$

Then $y_j \in PH^*$ and $\deg y_j \leq m + 2$. Also, the degree of b must be odd or else $Sq^n \sigma^* b =$

$\sigma^*(Sq^n b) = 0$. It may be assumed that $b \in PH^*$. Then the following formulas hold

$$
h_2(\sigma^* x) = h_2(\sum_j \gamma_j \sigma^* y_j) + h_2(Sq^n \sigma^* b)
$$

$$
= (\sum_j \gamma_j h_2(\sigma^* y_j)) + Sq^n h_2(\sigma^* b) \tag{3.22}
$$

$$
= \sum_j \gamma_j (\sigma^* \otimes \sigma^*) h_1(y_j)
$$

since $Sq^n h_2(\sigma^* b) = 0$ for degree reasons. Also, $h_1(y_j) \in H^p(X \wedge X)$, where $p \le m + 1$.

Then because X is 2-connected,

$$
\gamma_j (\sigma^* \otimes \sigma^*) h_1(y_j) \in B_m \otimes B_m. \tag{3.23}
$$

Hence $h_2(\sigma^* x) \in B_m \otimes B_m$ as desired. QED

The action of the Steenrod algebra on H^* has the property that decomposable elements are mapped to decomposables elements by Steenrod operations. Thus there is an induced action of the Steenrod algebra on QH^*.

PROPOSITION 3.5.   For all $s \ge 0$,

$$
QH^{d(r,2^s)-1} = \begin{cases} Sq^2 Sq^1 QH^{2^{s+3}} & \text{if } r = 2; \\ 0 & \text{if } r \ge 3. \end{cases}
$$

PROOF: Assume that the proposition is false and for a fixed value of s, choose the largest r such that the proposition is not satisfied. Pick an indecomposable element $x \in PH^{d(r,2^s)-1}$.

If $r = 2$ assume $x \notin \operatorname{im} Sq^2 Sq^1$ modulo decomposable elements. There is a factorization of the Steenrod algebra given by

$$Sq^{d(r,2^s)+1} = Sq^2 Sq^1 Sq^{d(r,2^s)-2} + Sq^{d(r,2^s)} Sq^1. \tag{3.24}$$

Consider the two stage Postnikov system

$$K(Z_2, d(r, 2^s) - 1, d(r+1, 2^s) - 4)$$

$$\downarrow j$$

$$\begin{array}{ccc} & E & \xrightarrow{\;v\;} & K(Z_2, d(r+1, 2^s) - 1) \\ \overset{\tilde{f}}{\diagup} & & \\ & \downarrow p & \end{array} \tag{3.25}$$

$$X \xrightarrow{\;f\;} K(Z_2, d(r, 2^s) - 1) \xrightarrow{\;g\;} K(Z_2, d(r, 2^s), d(r+1, 2^s) - 3)$$

where f represents x and $g^* = (\, Sq^1 \quad Sq^{d(r,2^s)-2} \,)$. The map v comes from the relation (3.24) above and corresponds to a secondary operation ϕ. The class v is primitive and satisfies $h_1(v) = p^* \iota_{d(r,2^s)-1} \otimes p^* \iota_{d(r,2^s)-1}$ and $j^*(v) = Sq^{d(r,2^s)} \iota_{d(r,2^s)-1} + Sq^2 Sq^1 \iota_{d(r+1,2^s)-4} = Sq^2 Sq^1 \iota_{d(r+1,2^s)-4}$.

By Lemma 3.2, $Sq^1 x = 0$. The degree of $Sq^{d(r,2^s)-2} x$ is $d(r+1, 2^s) - 3 = d(1, 2^{r+s} + 2^{r-1} - 1) - 1$. Since $2^{r+s} + 2^{r-1} - 1$ is not a power of two, Proposition 3.3 implies $Sq^{d(r,2^s)-2} x$ is decomposable. But $Sq^{d(r,2^s)-2} x$ is an odd degree primitive, so $Sq^{d(r,2^s)-2} x = 0$. Thus the lifting $\tilde{f}$ exists.

Because x is primitive, f is an H-map. If $\tilde{f}$ were an H-map, then $h_1(\phi(x))$ could be computed. Unfortunately, this is not necessarily the case; but it will be shown that Ωf is an h_1-map so that $h_2(\Omega\phi(x))$ can be computed. Since D_f, the H-deviation of f, is homotopic to $pD_{\tilde{f}}$, $D_{\tilde{f}}$ factors as a composition

$$X \wedge X \xrightarrow{D} K(Z_2, d(r, 2^s) - 1, d(r+1, 2^s) - 4) \xrightarrow{j} E. \tag{3.26}$$

The matrix $[D]$ satisfies the relation $(\Omega g)h_1(f) = (1 + T^*)[D]$, where $h_1(f)$ is the h_1-deviation of f and Ωg is given by the matrix

$$\Omega g = \begin{pmatrix} Sq^1 \\ Sq^{d(r,2^*)-2} \end{pmatrix}. \tag{3.27}$$

Thus

$$(1 + T^*)[D] = \begin{pmatrix} Sq^1 h_1(f) \\ Sq^{d(r,2^*)-2} h_1(f) \end{pmatrix}; \tag{3.28}$$

and since $h_1(f) \in H^{d(r,2^*)-2}(X \wedge X)$,

$$(1 + T^*)[D] \in \begin{pmatrix} Sq^1 H^{d(r,2^*)-2}(X \wedge X) \\ \xi H^*(X \wedge X) \end{pmatrix}. \tag{3.29}$$

The indeterminacy of $h_1(f)$ depends on the choice of H-structure of f. The value of $h_1(f)$ may be altered by any element contained in $(1 + T^*)H^{d(r,2^*)-2}(X \wedge X)$ by the appropriate change in H-structure for f. Thus the H-structure of f may be fixed so that

$$h_1(f) \in \bigoplus_{n=0}^{d(r-1,2^*)-4} (H^{d(r-1,2^*)-1+n} \otimes H^{d(r-1,2^*)-1-n}). \tag{3.30}$$

Then by the Cartan formula

$$Sq^1 h_1(f) \in \bigoplus_{n=0}^{d(r-1,2^s)-4} (Sq^1 H^{d(r-1,2^s)-1+n} \otimes H^{d(r-1,2^s)-1-n} +$$

$$H^{d(r-1,2^s)-1+n} \otimes Sq^1 H^{d(r-1,2^s)-1-n}).$$

(3.31)

Because $Sq^1 x = 0$, $Sq^1 h_1(f)$ must be contained in the indeterminacy, i.e. $Sq^1 h_1(f) \in$ im$(1 + T^*)$. If $Sq^1 h_1(f) \neq 0$, then for degree reasons

$$Sq^1 h_1(f) \in Sq^1 H^{d(r-1,2^s)-1} \otimes H^{d(r-1,2^s)-1} + H^{d(r-1,2^s)-1} \otimes Sq^1 H^{d(r-1,2^s)-1}.$$ (3.32)

But since $d(r - 1, 2^s)$ is not a power of two, Lemma 3.2 implies that $Sq^1 H^{d(r-1,2^s)-1} \in$ DH^*. Therefore $(1 + T^*)[D] \in H^* \otimes DH^* + DH^* \otimes H^*$, and hence $[D] \in H^* \otimes DH^* + DH^* \otimes$ $H^* + \ker(1 + T^*)$. There is a choice of basis for $\ker(1 + T^*)$ such that the basis elements have the form $a \otimes b + b \otimes a$ or $c \otimes c$. The matrix $[D]$ has bidegree $(d(r, 2^s) - 1, d(r+1, 2^s) - 4)$. There are no elements of the form $c \otimes c$ with total degree equal to $d(r, 2^s) - 1$. If $c \otimes c$ has total degree equal to $d(r + 1, 2^s) - 4$, then

$$\deg c = d(r, 2^s) - 2 = \begin{cases} d(1, 2^{s+1}) & \text{if } r = 2; \\ d(1, 2^{r+s-1} + \sum_{j=0}^{r-3} 2^j) & \text{if } r \geq 3. \end{cases}$$ (3.33)

In either case c must be decomposable by Proposition 2.6. Given a basis element $a \otimes b + b \otimes a$, suppose $\overline{\Delta}(a) = \sum_p a_p \otimes a'_p$ and $\overline{\Delta}(b) = \sum_q b_q \otimes b'_q$. Then

$$\overline{\Delta}(ab) = a \otimes b + b \otimes a + \sum_p (a_p b \otimes a'_p + a_p \otimes a'_p b) + \sum_q (ab_q \otimes b'_q + b_q \otimes ab'_q)$$

$$+ (\sum_p a_p \otimes a'_p)(\sum_q b_q \otimes b'_q),$$

(3.34)

so that $a \otimes b + b \otimes a \in H^* \otimes DH^* + DH^* \otimes H^* + \operatorname{im} \overline{\Delta}$. It may be concluded that

$[D] \in H^* \otimes DH^* + DH^* \otimes H^* + \operatorname{im} \overline{\Delta}$. The indeterminacy for $[D]$ arising from different

choices of the lifting $\tilde{f}$ is precisely equal to $\operatorname{im} \overline{\Delta}$. Therefore there exists a choice of $\tilde{f}$ such

that $[D] \in H^* \otimes DH^* + DH^* \otimes H^*$.

Now consider the looped operation $\Omega \phi$ corresponding to the Postnikov system

$$K(Z_2, d(r, 2^s) - 2, d(r+1, 2^s) - 5)$$

$$\downarrow \Omega j$$

$$\Omega E \xrightarrow{\Omega v} K(Z_2, d(r+1, 2^s) - 2)$$

$$\Omega \tilde{f} \nearrow \qquad \downarrow \Omega p \tag{3.35}$$

$$\Omega X \xrightarrow{\Omega f} K(Z_2, d(r, 2^s) - 2) \xrightarrow{\Omega g} K(Z_2, d(r, 2^s) - 1, d(r+1, 2^s) - 4)$$

The loop map Ωf, representing $\sigma^* x$, has h_1-deviation

$$h_1(\Omega f) = (\sigma^* \otimes \sigma^*)[D_f] = 0. \tag{3.36}$$

The lifting $\Omega \tilde{f}$ is an H-map whose h_1-deviation satisfies the formula $(\Omega^2 p) h_1(\Omega \tilde{f}) = h_1(\Omega f)$. Thus $h_1(\Omega \tilde{f})$ factors as a composition

$$\Omega X \wedge \Omega X \xrightarrow{h_1} K(Z_2, d(r, 2^s) - 3, d(r+1, 2^s) - 6) \xrightarrow{\Omega^2 j} \Omega^2 E. \tag{3.37}$$

But $h_1(\Omega \tilde{f})$ is homotopic to the adjoint of the composition

$$\Sigma \Omega X \wedge \Sigma \Omega X \xrightarrow{\epsilon \wedge \epsilon} X \wedge X \xrightarrow{D_f} E, \tag{3.38}$$

and $D_{\bar{f}}$ is equal to the composition

$$X \wedge X \xrightarrow{D} K(Z_2, d(r, 2^s) - 1, d(r + 1, 2^s) - 4) \xrightarrow{j} E. \qquad (3.39)$$

Hence h_1 is homotopic to the adjoint of the composition

$$\Sigma\Omega X \wedge \Sigma\Omega X \xrightarrow{\epsilon\wedge\epsilon} X \wedge X \xrightarrow{D} K(Z_2, d(r, 2^s) - 1, d(r + 1, 2^s) - 4); \qquad (3.40)$$

and so $[h_1] = (\sigma^* \otimes \sigma^*)[D] = 0$. This implies that $\Omega\tilde{f}$ is an h_1-map. Note also here that

since v is an H-map, Ωv is also an h_1-map. Under these circumstances it is possible to

compute $h_2(\Omega\phi(x))$.

The element $\sigma^* x$ is a non-zero primitive; thus it is either indecomposable or it is

the square of an element q of degree $d(r - 1, 2^s) - 1$. Since the degree of q is odd, q

may be written as a sum $w + d$, where $w \in PH^*(\Omega X)$ and $d \in DH^*(\Omega X)$. In the

case that $\sigma^* x = q^2 = w^2 + d^2$, q^2 must be indecomposable in $\xi H^*(\Omega X)$; hence $w^2 \neq 0$.

Since w is an odd degree primitive, $w = \sigma^* e$, and $w^2 = (\sigma^* e)^2 = Sq^{d(r-1,2^s)-1}(\sigma^* e) =$

$\sigma^*(Sq^1 Sq^{d(r-1,2^s)-2} e)$. Then $Sq^{d(r-1,2^s)-2} e$ must be indecomposable in order for w^2 to be

non-zero. However, the degree of $Sq^{d(r-1,2^s)-2} e$ is

$$d(r, 2^s) - 2 = \begin{cases} d(1, 2^{s+1}) & \text{if } r = 2, \\ d(1, 2^{r+s-1} + \sum_{j=0}^{r-3} 2^j) & \text{if } r \geq 3, \end{cases} \qquad (3.41)$$

and there are no indecomposables in those degrees by Proposition 2.6. Therefore $\sigma^* x$ must

be indecomposable.

Lemma 3.4 implies the existence of an $\mathcal{A}(2)$ sub-Hopf algebra B_m of $H^*(\Omega X)$ such that

$h_2(f) \in B_m \otimes B_m$ and an element $t \in PH_*(\Omega X)$ satisfying $\langle t, \sigma^* x \rangle \neq 0$ and $\langle t, B_m \rangle = 0$.

If $r = 2$, then $\sigma^* x \notin \operatorname{im} Sq^2 Sq^1$ modulo decomposable elements, so t may be chosen with

the property that $t Sq^2 Sq^1 = 0$. This is a consequence of the following argument. If

$\sigma^* x = Sq^2 Sq^1 a_1 + d_1$, where a_1 is indecomposable and d_1 decomposable, then since the

degree of a_1 is odd, $a_1 \in \operatorname{im} \sigma^*$ modulo decomposables. Therefore $\sigma^* x = Sq^2 Sq^1 \sigma^* a_2 + d_2$,

where d_2 is decomposable. But then $\sigma^*(x + Sq^2 Sq^1 a_2)$ is a decomposable of degree $2^{s+3} + 2$.

This means that $\sigma^*(x + Sq^2 Sq^1 a_2) = a_3^2 = Sq^{2^{s+2}+1} a_3$, and a_3 is an indecomposable of

degree $2^{s+2} + 1$. But then $a_3 = \sigma^* a_4 + d_3$, where a_4 is indecomposable in $H^{2^{s+2}+2}$

and d_3 is decomposable. The existence of such an a_4 is impossible by Proposition 2.6.

Therefore $\sigma^*(x + Sq^2 Sq^1 a_2) = 0$; but this is also impossible since $x \notin \operatorname{im} Sq^2 Sq^1$ modulo

decomposables. Hence t can be chosen with the desired properties.

Then

$$h_2(\Omega \phi(x)) = \sigma^* x \otimes \sigma^* x + \operatorname{im}(1 + T^*) + \operatorname{im} Sq^2 Sq^1 + H^*(\Omega X) \otimes B_m + B_m \otimes H^*(\Omega X), \quad (3.42)$$

and hence

$$\langle t \otimes t, h_2(\Omega \phi(x)) \rangle = \langle t \otimes t, \sigma^* x \otimes \sigma^* x + \operatorname{im}(1 + T^*) + \operatorname{im} Sq^2 Sq^1$$

$$+ H^*(\Omega X) \otimes B_m + B_m \otimes H^*(\Omega X) \rangle \quad (3.43)$$

$$= \langle t \otimes t, \sigma^* x \otimes \sigma^* x + \operatorname{im} Sq^2 Sq^1 \rangle.$$

Suppose $tSq^1 \neq 0$. Then since $tSq^1 \in PH_*(\Omega X)$, and the degree of tSq^1 is $d(r, 2^s) - 3$,

$QH^{d(r,2^s)-3}(\Omega X) \neq 0$. This implies that $PH^{d(r,2^s)-3}(\Omega X) \neq 0$, and hence that the image

of σ^* is non-zero in degree $d(r, 2^s) - 3$. But $QH^{d(r,2^s)-2} = 0$ by Proposition 2.6, so this

is impossible. Therefore $tSq^1 = 0$. Suppose for $r \geq 3$ that $tSq^2Sq^1 \neq 0$. The same

argument may be applied to the conclusion that $QH^{d(r,2^s)-4} \neq 0$. Again this is impossible

by Proposition 2.6. Therefore $tSq^2Sq^1 = 0$ no matter what the value of r.

The value of $(t \otimes t)Sq^2Sq^1$ is then equal to

$$(tSq^2 \otimes t + tSq^1 \otimes tSq^1 + t \otimes tSq^2)Sq^1$$

$$=tSq^2Sq^1 \otimes t + tSq^2 \otimes tSq^1 + tSq^1 \otimes tSq^2 + t \otimes tSq^2Sq^1 \tag{3.44}$$

$$=0.$$

Therefore

$$\langle t \otimes t, h_2(\Omega\phi(x)) \rangle = \langle t \otimes t, \sigma^*x \otimes \sigma^*x \rangle \neq 0. \tag{3.45}$$

This implies in particular that $\Omega\phi(x) \neq 0$, which means σ^* is non-zero on some indecom-

posable of H^* of degree $d(r + 1, 2^s) - 1$. But this is a contradiction; since by the choice of

r, $QH^{d(r+1,2^s)-1} = 0$. QED

The following proposition takes care of the indecomposables from the proposition

above in the case where $r = 2$ and $s \geq 1$.

PROPOSITION 3.6.   *For all $s \geq 1$, $QH^{d(2,2^s)-1} = 0$.*

PROOF: Suppose the proposition is false, and pick the smallest s such that the proposition

fails. Pick an indecomposable element $x \in PH^{d(2,2^s)-1}$; such an element exists since all odd

indecomposables have primitive representatives. There is a factorization of the Steenrod

algebra given by

$$Sq^{d(2,2^s)} = Sq^8(Sq^{2^{s+3}-5}Sq^1 + Sq^{2^{s+3}-6}Sq^2 + Sq^4 Sq^{2^{s+3}-8}) + Sq^{2^{s+3}}Sq^4. \qquad (3.46)$$

Consider the two stage Postnikov system

$$K(Z_2, n_1 - 1, n_2 - 1, n_3 - 1, n_4 - 1)$$

$$\downarrow j$$

$$\begin{array}{ccc} & E & \xrightarrow{v} & K(Z_2, d(3,2^s) - 2) \\ {}^{\tilde{f}}\nearrow & & \\ & & \downarrow p \\ X & \xrightarrow{f} & K(Z_2, d(2,2^s) - 1) & \xrightarrow{g} & K(Z_2, n_1, n_2, n_3, n_4) \end{array} \qquad (3.47)$$

where f represents x and $g^* = \left(Sq^1 \quad Sq^2 \quad Sq^4 \quad Sq^{2^{s+3}-8} \right)$. The (n_1, n_2, n_3, n_4) are

given by $(d(2,2^s), d(2,2^s) + 1, d(2,2^s) + 3, d(2,2^{s+1}) - 9)$. The map v comes from the

relation (3.46) above and corresponds to a secondary operation ϕ. The class v satisfies

$\overline{\Delta}(v) = p^* \iota_{d(2,2^s)-1} \otimes p^* \iota_{d(2,2^s)-1}$ and $j^*(v) = Sq^8(Sq^{2^{s+3}-5}\iota_{d(2,2^s)-1} + Sq^{2^{s+3}-6}\iota_{d(2,2^s)} +$

$Sq^4 \iota_{d(2,2^s)+2}) + Sq^{2^{s+3}}\iota_{d(2,2^{s+1})-10}.$

By Lemma 3.2, $Sq^1 x = 0$. The degree of $Sq^2 x$ is $d(2, 2^s) + 1 = d(1, 2^{s+1} + 1) - 1$.

By Proposition 3.3, $Sq^2 x = 0$. The degree of $Sq^4 x$ is $d(2, 2^s) + 3 = d(3, 2^{s-1}) - 1$. By

Proposition 3.5, $Sq^4 x = 0$. The degree of $Sq^{2^{s+3}-8} x$ is $d(2, 2^{s+1}) - 9 = d(2, \sum_{j=0}^s 2^j) - 1$.

Since $\sum_{j=0}^s 2^j$ is not a power of two, Proposition 3.3 implies $Sq^{2^{s+3}-8} x = 0$. Therefore,

the lifting $\tilde{f}$ exists.

The secondary operation ϕ associated to v satisfies the formula

$$\overline{\Delta}\phi(x) = x \otimes x + \operatorname{im}\left(Sq^8(Sq^{2^{s+3}-5} + Sq^{2^{s+3}-6} + Sq^4) + Sq^{2^{s+3}}\right). \tag{3.48}$$

Pick an element $t \in PH_*$ such that $\langle t, x \rangle \neq 0$. Then

$$\langle t^2, \phi(x) \rangle = \langle t \otimes t, \overline{\Delta}\phi(x) \rangle$$

$$= \langle t \otimes t, x \otimes x + \operatorname{im}\left(Sq^8(Sq^{2^{s+3}-5} + Sq^{2^{s+3}-6} + Sq^4) + Sq^{2^{s+3}}\right). \tag{3.49}$$

For reasons of degree and by the Cartan formula, $(t \otimes t)Sq^{2^{s+3}} = tSq^{2^{s+2}-1} \otimes$

$tSq^{2^{s+2}+1} + tSq^{2^{s+2}} \otimes tSq^{2^{s+2}} + tSq^{2^{s+2}+1} \otimes tSq^{2^{s+2}-1}$. But $tSq^{2^{s+2}+1}$ is a homology

primitive of degree $d(1, 2^s)$; by Proposition 2.6, $tSq^{2^{s+2}+1} = 0$. Thus $(t \otimes t)Sq^{2^{s+3}} =$

$tSq^{2^{s+2}} \otimes tSq^{2^{s+2}}$. Suppose that $tSq^{2^{s+2}} \neq 0$. Then there is an indecomposable element

$y \in H^{d(2, 2^{s-1})-1}$ such that $\langle tSq^{2^{s+2}}, y \rangle \neq 0$. By Proposition 3.5, there is an indecom-

posable element z such that $Sq^2 z = y$ modulo decomposable elements. This means that

$tSq^{2^{s+2}}Sq^2 \neq 0$. But

$$tSq^{2^{s+2}}Sq^2 = tSq^4 Sq^{2^{s+2}-2} + tSq^{2^{s+2}+2}$$

$$= tSq^4 Sq^{2^{s+2}-2}$$

(3.50)

$$= tSq^4 Sq^2 Sq^{2^{s+2}-4} + tSq^4 Sq^1 Sq^{2^{s+2}-4} Sq^1$$

$$= 0$$

since $tSq^4 Sq^2$ and $tSq^4 Sq^1$ are homology primitives of degree $d(1, \sum_{j=0}^{s} 2^j) - 1$ and $d(1, \sum_{j=0}^{s} 2^j)$, respectively (which are zero by Propositions 3.3 and 2.6). Thus it may be concluded that $(t \otimes t)Sq^{2^{s+3}} = 0$ and $tSq^{2^{s+2}} = 0$.

If $s \geq 2$, then $(t \otimes t)Sq^8 = tSq^4 \otimes tSq^4$ because tSq^8, tSq^7, tSq^6 and tSq^5 are homology primitives of degree $d(2, \sum_{j=0}^{s-1} 2^j) - 1$, $d(2, \sum_{j=0}^{s-1} 2^j)$, $d(1, \sum_{j=0}^{s} 2^j) - 1$ and $d(1, \sum_{j=0}^{s} 2^j)$, respectively. These are all zero by Propositions 3.3, 2.6, 3.3 and 2.6, respectively. In the case that $s = 1$, $(t \otimes t)Sq^8 = tSq^4 \otimes tSq^4$ for the same reasons (except that $tSq^8 = tSq^{2^{1+2}} = tSq^{2^{s+2}} = 0$ by the argument in the previous paragraph). For some α, $\beta \in \mathcal{A}(2)$, $Sq^{2^{s+3}-5} + Sq^{2^{s+3}-6} = Sq^1 \alpha + Sq^2 \beta$. Note that $tSq^4 Sq^4$, $tSq^4 Sq^3$, $tSq^4 Sq^2$ and $tSq^4 Sq^1$ are homology primitives of degree $d(2, \sum_{j=0}^{s-1} 2^j) - 1$, $d(2, \sum_{j=0}^{s-1} 2^j)$, $d(1, \sum_{j=0}^{s} 2^j) - 1$ and $d(1, \sum_{j=0}^{s} 2^j)$, respectively. These are all zero in the case that $s \geq 2$ by Propositions 3.3, 2.6, 3.3 and 2.6, respectively. In the case that $s = 1$, it is possible that $tSq^4 Sq^4 \neq 0$. However, $tSq^4 Sq^4 = tSq^7 Sq^1 + tSq^6 Sq^2$, and $t \in \ker Sq^7$ and $t \in \ker Sq^6$, so $tSq^4 Sq^4 = 0$.

Now by the Cartan formula

$$(t \otimes t)Sq^8(Sq^{2^{s+3}-5} + Sq^{2^{s+3}-6} + Sq^4) = (tSq^4 \otimes tSq^4)(Sq^1\alpha + Sq^2\beta + Sq^4)$$

(3.51)

$$= 0.$$

It has been shown that $(t \otimes t)(Sq^8(Sq^{2^{s+3}-5} + Sq^{2^{s+3}-6} + Sq^4) + Sq^{2^{s+3}}) = 0$. Therefore

$$\langle t^2, \phi(x) \rangle = \langle t \otimes t, x \otimes x \rangle \neq 0 \tag{3.52}$$

and hence $t^2 \neq 0$. But t^2 is a homology primitive of degree $d(1, 2^{s+2} + 1)$. This is a

contradiction to Proposition 2.6. QED

Up to this point it has been shown that QH^* is concentrated in degree 11, degrees

which are powers of two, or powers of two plus or minus one. In the next two lemmas it

is shown that the primitives in degrees which are powers of two plus one are connected

by Steenrod operations. This knowledge of the action of the Steenrod algebra on H^* is

essential in the later steps of the proof of the Main Theorem, for it enables us to show

$QH^n = 0$ in degrees $n \geq 17$ equal to a power of two plus one by showing $QH^{17} = 0$.

This will appear in section 4. It also enables us to show that certain cohomology classes

can be represented by h_1-maps. This is useful in the application of secondary operation

arguments.

LEMMA 3.7.   If $x \in PH^{d(1,2^s)-1}$ and either $s \geq 2$, or $s = 1$ and $Sq^2x = 0$, then

$x \in \operatorname{im} Sq^{2^{s+1}}$.

PROOF: Note that in the case that $s \geq 2$, $Sq^2 x$ is a primitive of degree $d(2, 2^{s-1}) - 1$, and

hence by Proposition 3.6, $Sq^2 x = 0$. Suppose the lemma is false; that is, there is an element

$x \in PH^{d(1,2^s)-1}$ for some $s \geq 1$ such that $Sq^2 x = 0$ and $x \notin \mathrm{im}\, Sq^{2^{s+1}}$. Then $x \notin \mathrm{im}\, Sq^{2^{s+1}}$

modulo decomposable elements. The reason is that if $x = Sq^{2^{s+1}} z + d_1$, where $d_1 \in DH^*$,

then $z = w + d_2$, where $w \in PH^*$ and $d_2 \in DH^*$, and hence $x = Sq^{2^{s+1}} w + d_3$, where

$d_3 \in DH^*$. But then $d_3 \in PH^*$, so $d_3 = 0$ since it has odd degree. This would give a

contradiction.

There is a factorization of the Steenrod algebra given by

$$Sq^{d(1,2^s)} = Sq^4 Sq^2 Sq^{2^{s+2}-4} + Sq^4 Sq^1 Sq^{2^{s+2}-4} Sq^1 + Sq^{2^{s+2}} Sq^2. \qquad (3.53)$$

Consider the two stage Postnikov system

$$K(Z_2, d(1,2^s) - 1, d(1,2^s), 2^{s+3} - 4)$$

$$\downarrow j$$

$$\qquad\qquad E \xrightarrow{\ v\ } K(Z_2, 2^{s+3} + 2) \qquad\qquad (3.54)$$

$$\qquad \tilde{f} \qquad\qquad\qquad \downarrow p$$

$$X \xrightarrow{\ f\ } K(Z_2, d(1,2^s) - 1) \xrightarrow{\ g\ } K(Z_2, d(1,2^s), d(1,2^s) + 1, 2^{s+3} - 3)$$

where f represents x, $g^* = \begin{pmatrix} Sq^1 & Sq^2 & Sq^{2^{s+2}-4} \end{pmatrix}$. The map v comes from the relation

(3.53) above and corresponds to a secondary operation ϕ. The class v satisfies the conditions that $\overline{\Delta}(v) = p^*\iota_{d(1,2^s)-1} \otimes p^*\iota_{d(1,2^s)-1}$ and $j^*(v) = Sq^4 Sq^1 Sq^{2^{s+2}-4}\iota_{d(1,2^s)-1} +$

$Sq^{2^{s+2}}\iota_{d(1,2^s)} + Sq^4 Sq^2 \iota_{2^{s+3}-4}$.

By Lemma 3.2, $Sq^1 x = 0$. The degree of $Sq^{2^{s+2}-4}x$ is $2^{s+3} - 3 = d(1, \sum_{j=0}^s 2^j) - 1$.

Since $\sum_{j=0}^s 2^j$ is not a power of two, Proposition 3.3 implies $Sq^{2^{s+2}-4}x = 0$. Therefore,

the lifting $\tilde{f}$ exists.

The secondary operation ϕ associated to v satisfies the formula

$$\overline{\Delta}\phi(x) = x \otimes x + \operatorname{im}(Sq^4 Sq^1 Sq^{2^{s+2}-4} + Sq^{2^{s+2}} + Sq^4 Sq^2). \tag{3.55}$$

Pick an element $t \in PH_*$ such that $\langle t, x \rangle \neq 0$. Since $x \notin \operatorname{im} Sq^{2^{s+1}}$ modulo decomposables,

t may be chosen so that $tSq^{2^{s+1}} = 0$. Then

$$\langle t^2, \phi(x) \rangle = \langle t \otimes t, \overline{\Delta}\phi(x) \rangle$$

$$= \langle t \otimes t, x \otimes x + \operatorname{im}(Sq^4 Sq^1 Sq^{2^{s+2}-4} + Sq^{2^{s+2}} + Sq^4 Sq^2). \tag{3.56}$$

Note that tSq^4, tSq^3 and $tSq^2 Sq^1$ are all zero since there are no homology primitives

in degree $d(1, \sum_{j=0}^{s-1} 2^j) - 1$ (by Proposition 3.3, except possibly if $s = 1$; but in this

case $tSq^4 = tSq^{2^{1+1}} = tSq^{2^{s+1}} = 0$ by the choice of t), or in degree $d(1, \sum_{j=0}^{s-1} 2^j)$ (by

Proposition 2.6). Then

$$(t \otimes t)(Sq^4(Sq^1 Sq^{2^{s+2}-4} + Sq^2) + Sq^{2^{s+2}}) = (tSq^2 \otimes tSq^2)(Sq^1 Sq^{2^{s+2}-4} + Sq^2)$$

$$= tSq^2 Sq^2 \otimes tSq^2 + tSq^2 \otimes tSq^2 Sq^2$$

$$\tag{3.57}$$

$$= tSq^3 Sq^1 \otimes tSq^2 + tSq^2 \otimes tSq^3 Sq^1$$

$$= 0;$$

and thus

$$\langle t^2, \phi(x) \rangle = \langle t \otimes t, x \otimes x \rangle \neq 0. \tag{3.58}$$

This implies that $t^2 \neq 0$. But t^2 is a homology primitive of degree $d(1, 2^{s+1})$. This is a

contradiction to Proposition 2.6. QED

LEMMA 3.8.   *If $x \in PH^9$, then $x \in \text{im } Sq^4$.*

PROOF: Note that in the case where $Sq^2 x = 0$, the lemma follows from Lemma 3.7.

Assume that the lemma is false, i.e. that there is a primitive element x of degree 9 such

that $Sq^2 x \neq 0$ and $x \notin \text{im } Sq^4$. Then a simple argument shows that $x \notin \text{im } Sq^4$ modulo

decomposables. There is a factorization of the Steenrod algebra given by

$$Sq^{11} = (Sq^4 Sq^3 + Sq^2 Sq^5) Sq^4 + Sq^8 Sq^3. \tag{3.59}$$

Consider the two stage Postnikov system

$$K(Z_2, 11, 12)$$

$$\downarrow j$$

$$\begin{array}{ccc} & E & \xrightarrow{v} & K(Z_2, 19) \end{array} \tag{3.60}$$

$$\tilde{f} \nearrow \quad \downarrow p$$

$$X \xrightarrow{f} K(Z_2, 9) \xrightarrow{g} K(Z_2, 12, 13)$$

where f represents x and $g^* = (\; Sq^3 \quad Sq^4\;)$. The map v comes from the relation (3.59) above and corresponds to a secondary operation ϕ. The class v is primitive and satisfies

$$h_1(v) = p^*\iota_9 \otimes p^*\iota_9 \text{ and } j^*(v) = Sq^8\iota_{11} + (Sq^4 Sq^3 + Sq^2 Sq^5)\iota_{12}.$$

The degree of $Sq^3 x$ is 12. There are no degree 12 primitives by Proposition 2.6 and Lemma 3.1. The degree of $Sq^4 x$ is 13. But $13 = d(1,3) - 1$; so Proposition 3.3 implies $Sq^4 x = 0$. Thus the lifting $\tilde{f}$ exists.

Because x is primitive, f is an H-map. If $\tilde{f}$ were an H-map, then $h_1(\phi(x))$ could be computed. Unfortunately, this is not necessarily the case; but it will be shown that Ωf is an h_1-map so that $h_2(\Omega\phi(x))$ can be computed. Since D_f, the H-deviation of f, is homotopic to $pD_{\tilde{f}}$, $D_{\tilde{f}}$ factors as a composition

$$X \wedge X \xrightarrow{D} K(Z_2, 11, 12) \xrightarrow{j} E. \tag{3.61}$$

The matrix $[D]$ satisfies the relation $(\Omega g)h_1(f) = (1 + T^*)[D]$, where $h_1(f)$ is the h_1-deviation of f and Ωg is given by the matrix

$$\Omega g = \begin{pmatrix} Sq^3 \\ Sq^4 \end{pmatrix}. \tag{3.62}$$

Thus

$$(1 + T^*)[D] = \begin{pmatrix} Sq^3 h_1(f) \\ Sq^4 h_1(f) \end{pmatrix}; \tag{3.63}$$

and since $h_1(f) \in H^8(X \wedge X)$, the H-structure of f may be chosen so that $h_1(f) \in$

$H^3 \otimes H^5 + H^4 \otimes H^4 = PH^3 \otimes PH^5 + PH^4 \otimes PH^4$. The following identities hold

$$Sq^3 PH^3 = 0, \qquad Sq^4 PH^3 = 0$$

$$Sq^2 PH^4 = 0, \qquad Sq^3 PH^4 = 0 \tag{3.64}$$

$$Sq^1 PH^5 = 0.$$

Using these relations along with the Cartan formula yields

$$(1 + T^*)[D] \in \left(\begin{array}{c} PH^3 \otimes PH^8 + PH^4 \otimes PH^7 \\ \\ PH^3 \otimes PH^9 + PH^4 \otimes PH^8 + PH^5 \otimes PH^7 \\ \\ + PH^4 \otimes \xi H^4 + \xi H^4 \otimes PH^4 \end{array} \right). \tag{3.65}$$

This implies that $(1 + T^*)[D] \in H^* \otimes DH^* + DH^* \otimes H^*$, and hence $[D] \in H^* \otimes DH^* +$

$DH^* \otimes H^* + \ker(1 + T^*)$. Arguing as in the proof of Proposition 3.5 (since $QH^6 = 0$), it

may be concluded that $[D] \in H^* \otimes DH^* + DH^* \otimes H^* + \operatorname{im} \overline{\Delta}$. The indeterminacy for $[D]$

arising from different choices of the lifting $\tilde{f}$ is precisely equal to $\operatorname{im} \overline{\Delta}$. Therefore there

exists a choice of $\tilde{f}$ such that $[D] \in H^* \otimes DH^* + DH^* \otimes H^*$.

Now consider the looped operation $\Omega \phi$ corresponding to the Postnikov system

$$K(Z_2, 10, 11)$$

$$\downarrow \Omega j$$

$$\begin{array}{ccc} & \Omega \tilde{f} \nearrow & E \xrightarrow{\Omega v} & K(Z_2, 18) \\ & & \downarrow \Omega p \end{array} \tag{3.66}$$

$$\Omega X \xrightarrow{\Omega f} K(Z_2, 8) \xrightarrow{\Omega g} K(Z_2, 11, 12)$$

The loop map Ωf, representing $\sigma^* x$, has h_1-deviation

$$h_1(\Omega f) = (\sigma^* \otimes \sigma^*)[D_f] = 0. \tag{3.67}$$

The lifting $\Omega \tilde{f}$ is an H-map which, as in the proof of Proposition 3.5, is an h_1-map. Note also here that since v is an H-map, Ωv is also an h_1-map. Under these circumstances it is possible to compute $h_2(\Omega \phi(x))$.

The element $\sigma^* x$ is a non-zero primitive. Since $Sq^2 x \neq 0$, and σ^* is non-zero on degree 11 primitives, $\sigma^*(Sq^2 x) = Sq^2 \sigma^* x \neq 0$. If $\sigma^* x$ is decomposable, then $Sq^2 \sigma^* x$ is decomposable; thus it must be the case that $Sq^2 \sigma^* x = y^2$ for some indecomposable y of degree 5. But $QH^6 = 0$, which implies $QH^5(\Omega X) = 0$. Therefore $\sigma^* x$ must be indecomposable.

Let S be the subspace of H^* spanned by elements of degree less than or equal to 5. Let B be the $\mathcal{A}(2)$ sub-Hopf algebra of $H^*(\Omega X)$ generated by $\sigma^*(S)$. Because

$$[h_2(\Omega f)] = (\sigma^* \otimes \sigma^*)[h_1(f)], \tag{3.68}$$

$[h_2(\Omega f)] \in B \otimes B$ by reasons of degree. Suppose $i : QB \to QH^*(\Omega X)$ and suppose $\widehat{\sigma^* x} \in \mathrm{im}\, i$. Then

$$\sigma^* x = \left(\sum_j \alpha_j \sigma^* z_j\right) + d, \tag{3.69}$$

where $\alpha_j \in \mathcal{A}(2)$, $\deg z_j \leq 5$ and $d \in DH^*(\Omega X)$. Note that $\alpha_j z_j$ must have degree 9. An easy calculation shows that in all cases, $\sigma^*(\alpha_j z_j)$ must be decomposable. Therefore $\sigma^* x$

is decomposable; but this is impossible. Therefore $\widehat{\sigma^* x} \notin \operatorname{im} i$. This implies the existence

of an element $t \in PH_*(\Omega X)$ satisfying $\langle t, \sigma^* x \rangle \neq 0$ and $\langle t, B \rangle = 0$. For degree reasons,

$tSq^4 = 0$. Then

$$h_2(\Omega \phi(x)) = \sigma^* x \otimes \sigma^* x + \operatorname{im}(1 + T^*) + \operatorname{im}(Sq^4 Sq^3 + Sq^2 Sq^5 + Sq^8)$$

$$(3.70)$$

$$+ H^*(\Omega X) \otimes B + B \otimes H^*(\Omega X),$$

and hence

$$\langle t \otimes t, h_2(\Omega \phi(x)) \rangle = \langle t \otimes t, \sigma^* x \otimes \sigma^* x + \operatorname{im}(Sq^4 Sq^3 + Sq^2 Sq^5 + Sq^8)$$

$$+ \operatorname{im}(1 + T^*) + H^*(\Omega X) \otimes B_m + B_m \otimes H^*(\Omega X) \rangle \qquad (3.71)$$

$$= \langle t \otimes t, \sigma^* x \otimes \sigma^* x + \operatorname{im}(Sq^4 Sq^3 + Sq^2 Sq^5 + Sq^8) \rangle.$$

Note the following identities

$$tSq^3 = 0, \qquad tSq^2 Sq^1 = 0, \qquad tSq^4 = 0, \qquad (3.72)$$

which follow from degree reasons and the fact that $QH^5(\Omega X) = 0$ because $QH^6 = 0$. Then

by the Cartan formula

$$(t \otimes t)Sq^4 Sq^3 = (tSq^2 \otimes tSq^2)Sq^3$$

$$= tSq^2 \otimes tSq^2 Sq^3 + tSq^2 Sq^3 \otimes tSq^2$$

$$(3.73)$$

$$= tSq^2 \otimes t(Sq^5 + Sq^4 Sq^1) + t(Sq^5 + Sq^4 Sq^1) \otimes tSq^2$$

$$= 0.$$

Also by the Cartan formula and the Adem relations

$$(t \otimes t)Sq^2 Sq^5 = (t \otimes t)Sq^6 Sq^1 = 0, \qquad (3.74)$$

and

$$(t \otimes t)Sq^8 = 0. \tag{3.75}$$

Therefore

$$\langle t \otimes t, h_2(\Omega\phi(x)) \rangle = \langle t \otimes t, \sigma^*x \otimes \sigma^*x \rangle \neq 0, \tag{3.76}$$

which implies that $\Omega\phi(x) \neq 0$. But $[\Omega\phi(x)] \in \operatorname{im} \sigma^*$ and the degree of $[\Omega\phi(x)]$ is 18. This

implies that $QH^{19} = QH^{d(2,2^1)-1} \neq 0$, which is a contradiction to Proposition 3.6. QED

PROPOSITION 3.9. $QH^{d(2,2^0)-1} = QH^{11} = 0$.

PROOF: Suppose x is a non-zero 11 dimensional indecomposable element.   Then by

Proposition 3.5, $x = Sq^2y$ modulo decomposable elements for some primitive element

y of degree 9. By Lemma 3.8, $y = Sq^4z$ for some $z \in PH^5$.   Then $Sq^2Sq^4z \neq 0$; but

$Sq^2Sq^4z = Sq^6z + Sq^5Sq^1z = 0$, which is a contradiction. QED

The string Steenrod operation connections from the previous lemmas is used in the

next lemma to show that certain cohomology classes can be represented by h_1-maps. There

is a slight abuse of notation; when it is stated that $h_1(f) = 0$, what it really means is that

$h_1(f) = 0$ for some choice of H structure map for f. This result is useful in computing

some of the operations that appear in later sections.

LEMMA 3.10.   If $x \in PH^{d(1,2^s)-1}$ for some $s \geq 0$, and f represents x, then $h_1(f) = 0$.

PROOF: If $s \geq 1$, then by Lemmas 3.7 and 3.8, $x \in \operatorname{im} Sq^{2^{s+1}}$. Suppose $x = Sq^{2^{s+1}} y$.

The element y is indecomposable and of odd degree. Therefore $y = w + d$, where w is

primitive and d decomposable. But $x + Sq^{2^{s+1}} d = Sq^{2^{s+1}} w$, so $Sq^{2^{s+1}} d$ is decomposable

and primitive. Since the degree of $Sq^{2^{s+1}} d$ is odd, $Sq^{2^{s+1}} d = 0$, and $x = Sq^{2^{s+1}} w$. If $s \geq 2$,

then the same argument may be repeated for w. In all cases, there is an element $z \in PH^5$

such that

$$x = Sq^{2^{s+1}} Sq^{2^s} \cdots Sq^4 z. \tag{3.77}$$

Let g represent z. Then

$$h_1(f) = Sq^{2^{s+1}} Sq^{2^s} \cdots Sq^4 h_1(g) = 0. \tag{3.78}$$

QED

§4 Further study of QH^*

In this section, the study of QH^* is continued. Up to this point it has been shown

that QH^* is concentrated in degrees which are powers of two, and powers of two plus or

minus one. The main results of this section are Propositions 4.5 and 4.9. In combination

with the results of the previous sections, Propositions 4.5 and 4.9 imply that the only

indecomposables of H^* must lie in degrees 3, 4, 5, 7, 8 or 9. The proofs of these propositions

depend heavily on several lemmas which precede them.

Proposition 4.5 states that there are no indecomposables in degrees greater than 15

which are powers of two or powers of two plus one. In order to prove this proposition,

several steps are needed. Since we have already shown that the indecomposables in degrees

$2^r + 1$ for $r \geq 2$ are connected by the Steenrod operations Sq^{2^r}, it is only necessary to

show $QH^{17} = 0$ to eliminate all the higher ones. This is done in Lemma 4.4. Also, in

order to eliminate the indecomposables in degrees 2^r for $r \geq 4$, we show that Sq^1 maps

these elements to indecomposables (Lemma 4.2). Since it will be shown that there are no

indecomposables in degrees $2^r + 1$, there can't be any in degrees 2^r. In order to prove

$QH^{17} = 0$, two more steps are needed. First, since a degree 17 primitive is in the image of

Sq^8Sq^4, we show that Sq^8Sq^4 factors as a sum $\alpha_i\phi_i$, where $\alpha_i \in \mathcal{A}(2)$ and ϕ_i is a secondary

operation. This is the content of Lemma 4.3. A more careful look shows that $\alpha_2 = Sq^1$ and

ϕ_2 applied to a five dimensional primitive must be a degree 16 indecomposable in order

to support an indecomposable in degree 17. However, in Lemma 4.1 it is shown that the

even degree primitives must be decomposable. This means that any even degree generator

is a non-primitive generator. A careful analysis of coproducts reveals that ϕ_2 cannot map

to an indecomposable, completing the proof.

Proposition 4.9 states that $QH^{2^r-1} = 0$ for all $r \geq 4$. The proof of this proposition

also requires several steps. We wish to construct a secondary operation associated to a

factorization of Sq^{2^r+1}. In order to do this, it is imperative to show that $Sq^1QH^{2^r-1} = 0$

for all $r \geq 4$. This is done in Lemma 4.8. We take an example of how the proof of this

lemma takes shape. Suppose $x \in PH^{15}$ and $Sq^1x \neq 0$. Then Sq^1x is either y^4 for some

$y \in PH^4$ or z^2 for some indecomposable z of degree 8. In the latter case, an implication

argument produces an indecomposable of degree 16. This is impossible. In the former

case, it is shown that any homology dual to y has non-zero square. We prove in Lemma

4.6 that under these circumstances, $Sq^1y = 0$. Now $y^4 = Sq^8Sq^4y$. Since $Sq^1y = 0$ and

$Sq^2y = 0$, there is a factorization $Sq^8Sq^4 = \beta_i\psi_i$, where $\beta_i \in \mathcal{A}(2)$ and ψ_i is a secondary operation. This factorization is used to prove $Sq^1x = y^4 = Sq^8Sq^4y = 0$.

LEMMA 4.1.   If $x \in PH^{2^r}$ and $r \geq 3$, then x is decomposable.

PROOF: Suppose the lemma is false and x is indecomposable. There is a factorization

$$Sq^{2^r+2} = Sq^4Sq^2Sq^{2^r-4} + Sq^4Sq^1Sq^{2^r-4}Sq^1 + Sq^{2^r}Sq^2. \qquad (4.1)$$

Consider the two stage Postnikov system

$$K(Z_2, 2^r + 1, 2^{r+1} - 5, 2^{r+1} - 4)$$

$$\downarrow j$$

$$\begin{matrix} & E & \xrightarrow{\ v\ } & K(Z_2, 2^{r+1} + 1) \\ \tilde{f} \nearrow & & & \\ & \downarrow p & & \end{matrix} \qquad (4.2)$$

$$X \xrightarrow{\ f\ } K(Z_2, 2^r) \xrightarrow{\ g\ } K(Z_2, 2^r + 2, 2^{r+1} - 4, 2^{r+1} - 3)$$

where f represents x and $g^* = (\, Sq^2 \quad Sq^{2^r-4} \quad Sq^{2^r-4}Sq^1 \,)$. The map v comes from the

relation (4.1) above and corresponds to a secondary operation ϕ. The class v is primitive

and satisfies $h_1(v) = p^*\iota_{2^r} \otimes p^*\iota_{2^r}$ and $j^*(v) = Sq^{2^r}\iota_{2^r+1} + Sq^4Sq^2\iota_{2^{r+1}-5} + Sq^4Sq^1\iota_{2^{r+1}-4}$.

The degree of Sq^2x is $2^r + 2 = d(1, 2^{r-2})$. Proposition 2.6 implies that Sq^2x is

decomposable. Therefore Sq^2x is the square of an odd degree class. Lemma 3.1 implies

$Sq^2x = 0$. The degree of $Sq^{2^r-4}x$ is $2^{r+1} - 4 = d(2, \sum_{j=0}^{r-3} 2^j)$. By Proposition 2.6, $Sq^{2^r-4}x$

is decomposable. Because $QH^{d(1,\sum_{j=0}^{r-3} 2^j)} = 0$, $Sq^{2^r-4}x$ must be the fourth power of an

odd degree class. Again Lemma 3.1 implies $Sq^{2^r-4}x = 0$. The degree of $Sq^{2^r-4}Sq^1x$ is

$2^{r+1} - 3 = d(1, \sum_{j=0}^{r-2} 2^j) - 1$. Since $r \geq 3$, $\sum_{j=0}^{r-2} 2^j$ is not a power of two, and Proposition

3.3 implies $Sq^{2^r-4}Sq^1x = 0$. Thus the lifting $\tilde{f}$ exists.

Because x is primitive, f is an H-map. If $\tilde{f}$ were an H-map, then $h_1(\phi(x))$ could

be computed. Unfortunately, this is not necessarily the case; but it will be shown that

Ωf is an h_1-map so that $h_2(\Omega\phi(x))$ can be computed. Since D_f, the H-deviation of f, is

homotopic to $pD_{\tilde{f}}$, $D_{\tilde{f}}$ factors as a composition

$$X \wedge X \overset{D}{\longrightarrow} K(Z_2, 2^r + 1, 2^{r+1} - 5, 2^{r+1} - 4) \overset{j}{\longrightarrow} E. \tag{4.3}$$

The matrix $[D]$ satisfies the relation $(\Omega g)h_1(f) = (1 + T^*)[D]$, where $h_1(f)$ is the h_1-

deviation of f and Ωg is given by the matrix

$$\Omega g = \begin{pmatrix} Sq^2 \\ Sq^{2^r-4} \\ Sq^{2^r-4}Sq^1 \end{pmatrix}. \tag{4.4}$$

Thus

$$(1 + T^*)[D] = \begin{pmatrix} Sq^2 h_1(f) \\ Sq^{2^r-4} h_1(f) \\ Sq^{2^r-4}Sq^1 h_1(f) \end{pmatrix}. \tag{4.5}$$

The indeterminacy of $h_1(f)$ depends on the choice of H-structure of f. The value of $h_1(f)$

may be altered by any element contained in $(1 + T^*)H^{2^r-1}(X \wedge X)$ by the appropriate change in H-structure for f. Thus the H-structure of f may be fixed so that $h_1(f)$ is possibly non-zero in

$$QH^{2^{r-1}-1} \otimes QH^{2^{r-1}} + H^* \otimes DH^* + DH^* \otimes H^*. \tag{4.6}$$

This is due to the fact that the only remaining indecomposables are in degrees which are a power of two, a power of two minus one or a power of two plus one. Note that since $Sq^1 x \in PH^{2^r+1}$, if h represents $Sq^1 x$, then $h_1(h) = 0$ by Lemma 3.10. This means that $Sq^1 h_1(f)$ is contained in the indeterminacy, $\mathrm{im}\,(1 + T^*)$. Therefore if $h_1(f) = \sum_j (y_j \otimes z_j) + d$, where y_j is an indecomposable of degree $2^{r-1} - 1$, z_j is an indecomposable of degree 2^{r-1} and $d \in H^* \otimes DH^* + DH^* \otimes H^*$, then $Sq^1 z_j$ must be decomposable in order for $Sq^1 h_1(f) \in \mathrm{im}\,(1 + T^*)$. Note that this fact implies that, as in the proof of Lemma 3.4, any non-trivial Steenrod operation applied to z_j must yield a decomposable. Then by the Cartan formula

$$Sq^2 h_1(f) \in H^{2^{r-1}+1} \otimes H^{2^r} + H^* \otimes DH^* + DH^* \otimes H^*,$$

$$Sq^{2^r-4} h_1(f) \in H^* \otimes DH^* + DH^* \otimes H^*, \tag{4.7}$$

$$Sq^{2^r-4} Sq^1 h_1(f) \in H^* \otimes DH^* + DH^* \otimes H^*,$$

and hence $(1 + T^*)[D] \in H^* \otimes DH^* + DH^* \otimes H^*$. Therefore $[D] \in H^* \otimes DH^* + DH^* \otimes H^* + \ker(1 + T^*)$. The matrix $[D]$ has components in total degree equal to $2^r + 1$, $2^{r+1} - 5$ and $2^{r+1} - 4$. Since $QH^{2^r-2} = QH^{d(1, \sum_{j=0}^{r-3} 2^j)} = 0$ by Proposition 2.6, the only diagonal terms of $[D]$ must lie in $H^* \otimes DH^* + DH^* \otimes H^*$. The argument from Proposition 3.5 may be applied, and the lifting $\tilde{f}$ may be altered so that $[D] \in H^* \otimes DH^* + DH^* \otimes H^*$. By looping the operation ϕ, the lifting $\Omega \tilde{f}$ is an h_1-map, and $h_2(\Omega \phi(x))$ may be computed.

The element $\sigma^* x$ is a non-zero primitive of odd degree. Therefore it is indecomposable. Lemma 3.4 may be applied so that there is an $\mathcal{A}(2)$ sub Hopf algebra B_m of $H^*(\Omega X)$ such that $h_2(\Omega f) \in B_m \otimes B_m$ and there is a homology primitive t such that $\langle t, \sigma^* x \rangle \neq 0$ and $\langle t, B_m \rangle = 0$.

Then

$$h_2(\Omega \phi(x)) = \sigma^* x \otimes \sigma^* x + \operatorname{im}(1 + T^*) + \operatorname{im}(Sq^{2^r} + Sq^4 Sq^2 + Sq^4 Sq^1)$$

$$\quad (4.8)$$

$$+ H^*(\Omega X) \otimes B_m + B_m \otimes H^*(\Omega X),$$

and hence

$$\langle t \otimes t, h_2(\Omega \phi(x)) \rangle = \langle t \otimes t, \sigma^* x \otimes \sigma^* x + \operatorname{im}(Sq^{2^r} + Sq^4 Sq^2 + Sq^4 Sq^1)$$

$$+ \operatorname{im}(1 + T^*) + H^*(\Omega X) \otimes B_m + B_m \otimes H^*(\Omega X) \rangle \quad (4.9)$$

$$= \langle t \otimes t, \sigma^* x \otimes \sigma^* x + \operatorname{im}(Sq^{2^r} + Sq^4 Sq^2 + Sq^4 Sq^1) \rangle.$$

Note the following relations

$$tSq^2 = 0, \qquad tSq^4 = 0. \quad (4.10)$$

$tSq^2 \in PH_{2^r-3}(\Omega X)$; thus tSq^2 is dual to a suspension element. But $QH^{2^r-2} = 0$ by Proposition 2.6, so $tSq^2 = 0$. A similar argument applies to tSq^4, unless $r = 3$, in which case $tSq^4 = 0$ for degree reasons.

The value of $(t \otimes t)Sq^4Sq^2$ is then equal to

$$(tSq^3 \otimes tSq^1 + tSq^1 \otimes tSq^3)Sq^2$$

$$= tSq^3Sq^2 \otimes tSq^1 + tSq^3Sq^1 \otimes tSq^1Sq^1 + tSq^3 \otimes tSq^1Sq^2$$

$$+ tSq^1Sq^2 \otimes tSq^3 + tSq^1Sq^1 \otimes tSq^3Sq^1 + tSq^1 \otimes tSq^3Sq^2 \qquad (4.11)$$

$$= 2(tSq^3 \otimes tSq^3)$$

$$= 0;$$

and $(t \otimes t)Sq^4Sq^1$ is equal to

$$(tSq^3 \otimes tSq^1 + tSq^1 \otimes tSq^3)Sq^1$$

$$= tSq^3Sq^1 \otimes tSq^1 + tSq^3 \otimes tSq^1Sq^1$$

$$+ tSq^1Sq^1 \otimes tSq^3 + tSq^1 \otimes tSq^3Sq^1 \qquad (4.12)$$

$$= tSq^2Sq^2 \otimes tSq^1 + tSq^1 \otimes tSq^2Sq^2$$

$$= 0.$$

Also, $(t \otimes t)Sq^{2^r} = 0$ for degree reasons. Therefore

$$\langle t \otimes t, h_2(\Omega\phi(x)) \rangle = \langle t \otimes t, \sigma^*x \otimes \sigma^*x \rangle \neq 0. \qquad (4.13)$$

This implies that $h_2(\Omega\phi(x)) \neq 0$ and, in particular, that $\Omega\phi(x) \neq 0$, which means σ^* is non-zero on some indecomposable of H^* of degree $2^{r+1} + 1$. So there is an element $z \in H^{2^{r+1}+1}$ such that $\sigma^*z = [\Omega\phi(x)]$. But by Lemma 3.10, if h represents z, $h_1(h) = 0$. This would imply that $h_2(\Omega\phi(x)) = 0$. This is a contradiction. QED

LEMMA 4.2.   *Let r be the largest integer such that $QH^{2^r} \neq 0$ and suppose $x \in H^{2^r}$ is indecomposable. Then $Sq^1 x$ is indecomposable.*

PROOF: Suppose that the lemma is false and that $Sq^1 x$ is decomposable. There is a factorization

$$Sq^{2^r+1} = Sq^2 Sq^1 Sq^{2^r-2} + Sq^{2^r} Sq^1. \qquad (4.14)$$

Let B denote the $\mathcal{A}(2)$ sub-Hopf algebra of H^* generated by elements whose degree is less than 2^r. Suppose there is an element y and a non-trivial Steenrod operation α such that $\alpha y = x$. Then y must be indecomposable. The degree of y must be odd since the only even degree indecomposables are in degrees of a power of two. Then $y = w + d$, where $w \in PH^*$ and $d \in DH^*$. So $x = \alpha y = \alpha(w + d) \in DH^*$ because Lemma 4.1 implies that the primitives in degree 2^r must be decomposable. In order to avoid a contradiction, no such y can exist; it may be concluded that $\hat{x}$ is not in the image of the map $QB \to QH^*$ induced by inclusion. This means there is an element $t \in PH_*$ such that $\langle t, x \rangle \neq 0$ and $\langle t, B \rangle = 0$. The element t has the property that $t\alpha = 0$ for all non-trivial $\alpha \in \mathcal{A}(2)$.

It was assumed that $Sq^1 x$ is decomposable. For degree reasons, $Sq^1 x \in \overline{B} \cdot \overline{B}$. The degree of $Sq^{2^r-2} x$ is $2^{r+1} - 2 = d(1, \sum_{j=0}^{r-2})$; thus $Sq^{2^r-2} x$ is also decomposable. Let A denote the sub-Hopf algebra of B generated by elements whose degree is less than or equal to $2^{r+1} - 2$. Then $\{Sq^1 Sq^{2^r-2} x\}$ is primitive and decomposable in $H^* /\!/ A$. Since its degree is odd, $\{Sq^1 Sq^{2^r-2} x\} = 0$. Also $\overline{\Delta} Sq^1 Sq^{2^r-2} x \in A \otimes A$. The argument from Lemma 2.5

may be applied to show that $Sq^1 Sq^{2^r-2} x \in \overline{B} \cdot \overline{B}$. Note here also that for degree reasons

$\overline{\Delta}(x) \in B \otimes B$.

Under these conditions the secondary operation found in Lin [9] may be applied. The

resulting formula is

$$\overline{\Delta}\phi(x) = x \otimes x + \mathrm{im}\,(Sq^2 + Sq^{2^r}) + H^* \otimes B + B \otimes H^*. \tag{4.15}$$

Then

$$\langle t^2, \phi(x) \rangle = \langle t \otimes t, \overline{\Delta}\phi(x) \rangle$$

$$= \langle t \otimes t, x \otimes x + \mathrm{im}\,(Sq^2 + Sq^{2^r}) + H^* \otimes B + B \otimes H^* \rangle \tag{4.16}$$

$$= \langle t \otimes t, x \otimes x \rangle$$

$$\neq 0.$$

This means that $0 \neq t^2 \in PH_{2^{r+1}}$, which is a contradiction to the hypothesis concerning

r. QED

LEMMA 4.3.   *On cohomology classes whose degree is less than or equal to six and in the*

kernel of Sq^1, Sq^5 and $Sq^4 Sq^2$ there is a factorization

$$Sq^8 Sq^4 = Sq^4 \phi_1 + Sq^1 \phi_2 + Sq^6 \phi_3 + Sq^7 \phi_4, \tag{4.17}$$

where the ϕ_i are given by the matrix relation

$$\begin{matrix} \phi_1 \\ \phi_2 \\ \phi_3 \\ \phi_4 \end{matrix} \begin{bmatrix} Sq^8 & Sq^4 & Sq^3 \\ Sq^8 Sq^3 & 0 & Sq^4 Sq^2 \\ Sq^6 & Sq^2 & 0 \\ 0 & Sq^1 & 0 \end{bmatrix} \begin{bmatrix} Sq^1 \\ Sq^5 \\ Sq^4 Sq^2 \end{bmatrix} = \begin{bmatrix} Sq^9 \\ Sq^{10} Sq^2 \\ 0 \\ 0 \end{bmatrix}. \tag{4.18}$$

Similarly, on classes whose degree is less than or equal to seven and in the kernel of Sq^1

and Sq^2 there is a factorization

$$Sq^8 Sq^4 = Sq^4 \psi_1 + (Sq^9 + Sq^7 Sq^2)\psi_2 + Sq^6 Sq^5 \psi_3, \tag{4.19}$$

where the ψ_i are given by the matrix relation

$$\begin{matrix} \psi_1 \\ \psi_2 \\ \psi_3 \end{matrix} \begin{bmatrix} Sq^8 + Sq^4 Sq^4 & Sq^4 Sq^2 Sq^1 + Sq^7 \\ Sq^3 & Sq^2 \\ Sq^1 & 0 \end{bmatrix} \begin{bmatrix} Sq^1 \\ Sq^2 \end{bmatrix} = \begin{bmatrix} Sq^9 \\ 0 \\ 0 \end{bmatrix}. \tag{4.20}$$

These factorizations hold modulo the total indeterminacy of the right hand sides of the

equations (4.17) and (4.19).

PROOF: The proof that follows is for the first factorization. The proof for the second

factorization is similar. It should be noted here that the second factorization is due to Lin

and Williams [12]. Consider the Postnikov system

$$K(Z_2, n, n+4, n+5)$$

$$\downarrow j$$

$$E \xrightarrow{v} K(Z_2, n+8, n+11, n+6, n+5) \tag{4.21}$$

$$\downarrow p$$

$$K(Z_2, n) \xrightarrow{g} K(Z_2, n+1, n+5, n+6)$$

where $n \leq 6$, $g^* = (\begin{matrix} Sq^1 & Sq^5 & Sq^4Sq^2 \end{matrix})$, and $[v] = (v_i)$, where v_i represents the operation

ϕ_i and comes from the matrix relation (4.18) above. Because the excess of the operations

Sq^9 and $Sq^{10}Sq^2$ is greater than $n + 1$, the v_i represent primitive elements. Furthermore,

$$j^*v_1 = Sq^4\iota_{n+4} + Sq^3\iota_{n+5}$$

$$j^*v_2 = Sq^8Sq^3\iota_n + Sq^4Sq^2\iota_{n+5}$$

$$j^*v_3 = Sq^6\iota_n + Sq^2\iota_{n+4} \tag{4.22}$$

$$j^*v_4 = Sq^1\iota_{n+4}.$$

A simple calculation shows that

$$j^*(Sq^4v_1 + Sq^1v_2 + Sq^6v_3 + Sq^7v_4) = 0. \tag{4.23}$$

Since $Sq^4v_1 + Sq^1v_2 + Sq^6v_3 + Sq^7v_4$ is primitive,

$$Sq^4v_1 + Sq^1v_2 + Sq^6v_3 + Sq^7v_4 = \varepsilon p^*(Sq^8Sq^4\iota_n), \tag{4.24}$$

where $\varepsilon \in Z_2$. This is because the only $n + 12$ degree primitive of $H^*(K(Z_2, n); Z_2)$ which

is not in the kernel of p^* is $Sq^8Sq^4\iota_n$. It is not immediate that $Sq^9Sq^3\iota_n \in \ker p^*$, but

$Sq^9Sq^3 = Sq^7Sq^5$ and $Sq^5\iota_n \in \ker p^*$.

The coefficient ε will now be determined. Consider the map $f : K(Z, 2) \to K(Z_2, 4)$

given by $f^*\iota_4 = \iota_2^2$. There is a diagram

$$K(Z_2, 4, 8, 9)$$

$$\downarrow j$$

$$\tilde{f} \qquad E \xrightarrow{\ v\ } K(Z_2, 12, 15, 10, 9) \tag{4.25}$$

$$\nearrow \qquad \downarrow p$$

$$K(Z, 2) \xrightarrow{\ f\ } K(Z_2, 4) \xrightarrow{\ g\ } K(Z_2, 5, 9, 10)$$

Since ι_2^2 is primitive, f is an H-map and the H-deviation of $\tilde{f}$ factors as a composition

$$K(Z, 2) \wedge K(Z, 2) \xrightarrow{\ D\ } K(Z_2, 4, 8, 9) \xrightarrow{\ j\ } E. \tag{4.26}$$

Let $P_2(K(Z, 2))$ denote the projective plane of $K(Z, 2)$. There is an exact triangle (see Browder and Thomas [2])

$$\overline{H}^*(K(Z, 2)) \otimes \overline{H}^*(K(Z, 2))$$

$$\overline{\Delta} \nearrow \qquad \qquad \searrow \lambda \tag{4.27}$$

$$H^*(K(Z, 2)) \xleftarrow{\ h\ } H^*(P_2(K(Z, 2)))$$

Since the class $\iota_2^2 \in \ker \overline{\Delta}$, $\iota_2^2 = h(u_5)$. The exact triangle has the property that $\lambda(\iota_2^2 \otimes \iota_2^2) = u_5^2$. Because $\iota_2^2 \otimes \iota_2^2 \notin \operatorname{im} \overline{\Delta}$, $u_5^2 \neq 0$. But

$$u_5^2 = Sq^5 u_5 = Sq^2 Sq^1 Sq^2 u_5 + Sq^4 Sq^1 u_5 = Sq^4 Sq^1 u_5. \tag{4.28}$$

since $Sq^2 u_5 \in H^7(P_2(K(Z,2))) = 0$. Suppose $\iota_2 = h(u_3)$. Then since $Sq^1 u_5 \neq 0$, it must

be the case that $Sq^1 u_5 = u_3^2$. So

$$\begin{pmatrix} Sq^1 \\ Sq^5 \\ Sq^4 Sq^2 \end{pmatrix} (u_5) = \begin{pmatrix} u_3^2 \\ u_5^2 \\ 0 \end{pmatrix}. \tag{4.29}$$

By Williams [17],

$$[D] = \begin{pmatrix} \iota_2 \otimes \iota_2 \\ \iota_2^2 \otimes \iota_2^2 \\ 0 \end{pmatrix}. \tag{4.30}$$

Since the v_i are primitive,

$$\overline{\Delta}\phi_1(\iota_2^2) = (\, Sq^8 \quad Sq^4 \quad Sq^3 \,)[D] = \iota_2^4 \otimes \iota_2^2 + \iota_2^2 \otimes \iota_2^4$$

$$\overline{\Delta}\phi_2(\iota_2^2) = (\, Sq^8 Sq^3 \quad 0 \quad Sq^4 Sq^2 \,)[D] = 0$$

$$\tag{4.31}$$

$$\overline{\Delta}\phi_3(\iota_2^2) = (\, Sq^6 \quad Sq^2 \quad 0 \,)[D] = 0$$

$$\overline{\Delta}\phi_4(\iota_2^2) = (\, 0 \quad Sq^1 \quad 0 \,)[D] = 0.$$

It is not difficult to see that $\phi_1(\iota_2^2) = \iota_2^6$ and $\phi_i(\iota_2^2) = 0$ for $i = 2, 3, 4$. Then

$$(Sq^4 \phi_1 + Sq^1 \phi_2 + Sq^6 \phi_3 + Sq^7 \phi_4)(\iota_2^2) = Sq^4(\iota_2^6) = \iota_2^8 = Sq^8 Sq^4(\iota_2^2). \tag{4.32}$$

This implies that the coefficient ε must be equal to 1. QED

LEMMA 4.4. $QH^{17} = 0$.

PROOF: Suppose not and choose a non-zero $x \in PH^{17}$. Then by Lemmas 3.7 and 3.8,

$x = Sq^8 Sq^4 z$ for some $z \in PH^5$. $Sq^1 z = 0$ by Lemma 3.2, $Sq^5 z = z^2 = 0$ by Lemma 3.1

and $Sq^4 Sq^2 z \in PH^{11} = 0$ by Proposition 3.9. Thus by Lemma 4.3,

$$x = Sq^8 Sq^4 z = (Sq^4 \phi_1 + Sq^1 \phi_2 + Sq^6 \phi_3 + Sq^7 \phi_4) z. \tag{4.33}$$

The degree of $\phi_1(z)$ is 13; therefore $\phi_1(z) \in DH^*$ by Proposition 3.5. The degree of

$\phi_3(z)$ is 11; thus $\phi_3(z) \in DH^*$ by Proposition 3.9. The degree of $\phi_4(z)$ is 10; therefore

$\phi_4(z) \in DH^*$ by Proposition 2.6. Since x is indecomposable, and Steenrod operations map

decomposables to decomposables, it must be the case that $\phi_2(z)$ is indecomposable.

Consider the Postnikov system associated to $\phi_2(z)$

$$
\begin{array}{ccc}
 & K(Z_2, 5, 10) & \\
 & \downarrow j & \\
\bar{f} \quad & E \xrightarrow{v_2} K(Z_2, 16) & \tag{4.34} \\
\diagup & \downarrow p & \\
X \xrightarrow{f} K(Z_2, 5) & \xrightarrow{g} K(Z_2, 6, 11) &
\end{array}
$$

where $f^* \iota_5 = z$, $g = (\, Sq^1 \quad Sq^4 Sq^2 \,)$ and $j^* v_2 = Sq^8 Sq^3 \iota_5 + Sq^4 Sq^2 \iota_{10}$. Since f is an

H-map, the H-deviation of $\tilde{f}$ factors as a composition

$$X \wedge X \xrightarrow{D} K(Z_2, 5, 10) \xrightarrow{j} E, \tag{4.35}$$

where $[D]$ satisfies the formula

$$(1 + T^*)[D] = (\Omega g)h_1(f) = 0. \tag{4.36}$$

Then for degree reasons, $[D]$ satisfies the formula

$$[D] \in \left(\frac{\mathrm{im}\,(1 + T^*)}{PH^5 \otimes PH^5 + \mathrm{im}\,(1 + T^*)} \right). \tag{4.37}$$

By an analysis using the Cartan formula, Steenrod operations map elements in $\mathrm{im}\,(1 + T^*)$

into $\mathrm{im}\,(1 + T^*)$. Since $[v_2]$ is primitive, the operation ϕ_2 satisfies the formula

$$\overline{\Delta}\phi_2(z) = (\, Sq^8 Sq^3 \quad Sq^4 Sq^2 \,)[D], \tag{4.38}$$

which implies that $\overline{\Delta}\phi_2(z) \in Sq^4 Sq^2((PH^5 \otimes PH^5) + \mathrm{im}\,(1 + T^*))$. Let $a \in PH^5$. Then

$$Sq^4 Sq^2(a \otimes a) = Sq^4(Sq^2 a \otimes a + a \otimes Sq^2 a$$

$$\tag{4.39}$$

$$= Sq^2 a \otimes Sq^4 a + Sq^4 a \otimes Sq^2 a$$

Therefore $Sq^4 Sq^2(a \otimes a) \in \mathrm{im}\,(1 + T^*)$, and hence $\overline{\Delta}\phi_2(z) \in \mathrm{im}\,(1 + T^*)$. Because $\phi_2(z)$

must be an indecomposable of degree 16, Lemma 4.1 implies that $\phi_2(z)$ must correspond to

a non-primitive generator. Therefore there is an element $t \in H_8$ such that $\langle t^2, \phi_2(z) \rangle \neq 0$.

But

$$\langle t^2, \phi_2(z) \rangle = \langle t \otimes t, \overline{\Delta} \phi_2(z) \rangle$$

$$= \langle t \otimes t, \mathrm{im}\,(1 + T^*) \rangle \qquad (4.40)$$

$$= 0.$$

This is a contradiction. QED

PROPOSITION 4.5.   For all $r \geq 4$, $QH^{2^r} = 0$, and for all $s \geq 2$, $QH^{d(1,2^s)-1} = 0$.

PROOF: Choose the largest s such that $QH^{d(1,2^s)-1} \neq 0$ and the largest r such that $QH^{2^r} \neq 0$. Assume that either $r \geq 4$ or $s \geq 2$. If $r \geq 4$ then by Lemma 4.2, $QH^{2^r+1} = QH^{d(1,2^{r-2})-1} \neq 0$. Therefore $s \geq r - 2 \geq 2$, so it suffices to show that there is a contradiction if $s \geq 2$. Let x be a non-zero primitive of degree $d(1,2^s) - 1 = 2^{s+2} + 1$. Then arguing as in the proof of Lemma 3.10, $x = Sq^{2^{s+1}} Sq^{2^s} \cdots Sq^4 z$ for some $z \in PH^5$. But $Sq^8 Sq^4 z$ is an indecomposable of degree $17 = d(1,2^2) - 1$. By Lemma 4.4, $Sq^8 Sq^4 z = 0$; but the assumption that $s \geq 2$ implies $Sq^8 Sq^4 z \neq 0$. This is a contradiction. QED

LEMMA 4.6.   Suppose $x \in PH^4$ and that $x \otimes x \in \mathrm{im}\,\overline{\Delta}$. Then $Sq^1 x = 0$.

PROOF: Suppose that $Sq^1 x = z \neq 0$. Since x and z are primitives, there are corresponding elements in $H^*(P_2 X)$, u_5 and u_6. Note that $Sq^1 u_5 = u_6$. Since $x \otimes x \in \mathrm{im}\,\overline{\Delta}$, the exact sequence

$$\overline{H}^* \otimes \overline{H}^*$$

$$(4.41)$$

$$H^* \xleftarrow{\ h\ } H^*(P_2X)$$

implies that $\lambda(x \otimes x) = 0$. But by Browder and Thomas [**2**], $\lambda(x \otimes x) = u_5^2$. Therefore

$$0 = u_5^2$$

$$= Sq^5 u_5$$

$$= Sq^2 Sq^1 Sq^2 u_5 + Sq^4 Sq^1 u_5 \qquad (4.42)$$

$$= Sq^4 Sq^1 u_5$$

$$= Sq^4 u_6,$$

since $Sq^2 u_5 = 0$ because $Sq^2 x = 0$ and the image of λ in degree 7 is 0. There are

no elements of H^* with reduced coproduct equal to $z \otimes z$ since $QH^{10} = 0$. Therefore

$u_6^2 = \lambda(z \otimes z) \neq 0$. However

$$u_6^2 = Sq^6 u_6$$

$$= Sq^2 Sq^4 u_6 + Sq^5 Sq^1 u_6$$

$$\qquad (4.43)$$

$$= Sq^5 Sq^1 Sq^1 u_5$$

$$= 0.$$

This is a contradiction. QED

LEMMA 4.7.   If $x \in PH^4$ and $Sq^1 x = 0$, then $x^4 = 0$.

PROOF: $Sq^2 x \in PH^6 = 0$ and $Sq^1 x = 0$ by assumption. Thus the factorization

$$Sq^8 Sq^4 = Sq^4 \psi_1 + (Sq^9 + Sq^7 Sq^2)\psi_2 + Sq^6 Sq^5 \psi_3, \tag{4.44}$$

where the ψ_i are given by the matrix relation

$$\begin{matrix} \psi_1 \\ \psi_2 \\ \psi_3 \end{matrix} \begin{bmatrix} Sq^8 + Sq^4 Sq^4 & Sq^4 Sq^2 Sq^1 + Sq^7 \\ Sq^3 & Sq^2 \\ Sq^1 & 0 \end{bmatrix} \begin{bmatrix} Sq^1 \\ Sq^2 \end{bmatrix} = \begin{bmatrix} Sq^9 \\ 0 \\ 0 \end{bmatrix}, \tag{4.45}$$

can be applied to x. Note that $x^4 = Sq^8 Sq^4 x$ since the degree of x is 4. Consider the

diagram

$$K(Z_2, 4, 5)$$

$$\downarrow j$$

$$\tilde{f} \qquad E \xrightarrow{v} K(Z_2, 12, 7, 5) \tag{4.46}$$

$$\nearrow \qquad \downarrow p$$

$$X \xrightarrow{f} K(Z_2, 4) \xrightarrow{g} K(Z_2, 5, 6)$$

where $f^*(\iota_4) = x$, $g = (Sq^1 \quad Sq^2)$, $v = (v_i)$ and v_i represents the operation ψ_i. Since f

is an H-map, the H-deviation of $\tilde{f}$ factors as a composition

$$X \wedge X \xrightarrow{D} K(Z_2, 4, 5) \xrightarrow{j} E, \tag{4.47}$$

and $[D] = 0$ for degree reasons. Hence $\tilde{f}$ is an H-map. The v_i represent primitive

elements, so v is also an H-map. Therefore $\psi_i(x) \in PH^*$ for $i = 1, 2, 3$. $\psi_1(x) \in$

$PH^{12} = 0$. $\psi_2(x) \in PH^7$, which implies $Sq^9 \psi_2(x) = 0$ for degree reasons. Also,

$Sq^7 Sq^2 \psi_2(x) = Sq^3 Sq^4 Sq^2 \psi_2(x)$ and $Sq^4 Sq^2 \psi_2(x) \in PH^{13} = 0$. $Sq^5 \psi_3(x) \in PH^{10} = 0$,

so $Sq^6 Sq^5 \psi_3(x) = 0$. It is now clear that $Sq^8 Sq^4 x = x^4 = 0$. QED

The proof of the next lemma needs some preliminary explanation. We wish to show

$Sq^1 PH^{2^r-1} = 0$. Consider the following example as motivation. Suppose we have a

primitive x of degree 15 and $Sq^1 x = y \neq 0$. Then it is not difficult to see that either

$y = z^4$ for some $z \in H^4$, or $y = w^2$ for some indecomposable $w \in H^8$. Since in this

situation $y^2 = 0$, we can find an element a such that $\overline{\Delta}(a) = y \otimes y +$ other terms we wish

to ignore. Since the degree of a is 32, a must be decomposable. If $y = w^2$, then there

should be an element b such that $b^2 = a$ and $\overline{\Delta}(b) = w \otimes w +$ other terms. In this case b

is indecomposable and has degree 16. This is not possible. Thus $y = z^4$ and there should

be an element c such that $c^4 = a$ and $\overline{\Delta}(c) = z \otimes z +$ other terms. The existence of such

a c makes $Sq^1(z +$ other terms of less height$) = 0$, and hence $z^4 = 0$ by the two previous

lemmas. This would give a contradiction. The arguments given here are sloppy, but the

general idea of the proof is evident.

LEMMA 4.8.   For all $r \geq 4$, $Sq^1 PH^{2^r-1} = 0$.

PROOF: Let $x \in PH^{2^r-1}$ and suppose $Sq^1 x = y \neq 0$. By Lemma 4.1, y is decomposable.

An argument such as the one found in the proof of Lemma 3.2 may be applied to show

that $y = z^{2^k}$ for some indecomposable element $z \in H^{2^{r-k}}$. By the argument found in the

proof of Lemma 2.2 either $y^2 \neq 0$, or if $\langle \bar{y}, y \rangle \neq 0$ then $\bar{y}^2 \neq 0$.

Suppose $y^2 \neq 0$. Then $Sq^{2^r} Sq^1 x = Sq^{2^r} y = y^2 \neq 0$. On the other hand, $Sq^{2^r} Sq^1 x =$

$Sq^2 Sq^1 Sq^{2^r-2} x = 0$, since $PH^{2^{r+1}-2} = 0$. Therefore it must be the case that $y^2 = 0$.

Choose a Borel decomposition for H^*; as algebras

$$H^* \cong \frac{Z_2[x_1]}{(\varepsilon_1 x_1^{2^{r_1}})} \otimes \cdots \otimes \frac{Z_2[x_n]}{(\varepsilon_n x_n^{2^{r_n}})} \otimes A, \qquad (4.48)$$

where $\deg x_i$ is either 4 or 8, $r_i \geq 1$, A is an algebra which is generated by odd degree

elements and $\varepsilon_i \in Z_2$. If $\varepsilon_i = 0$, x_i has infinite height; if $\varepsilon_i = 0$, x_i is truncated at height

2^{r_i}. There is a dual decomposition, as coalgebras

$$H_* \cong \frac{\Gamma[\bar{x}_1]}{(\varepsilon_1 \gamma_{2^{r_1}}(\bar{x}_1))} \otimes \cdots \otimes \frac{\Gamma[\bar{x}_n]}{(\varepsilon_n \gamma_{2^{r_n}}(\bar{x}_n))} \otimes A^*, \qquad (4.49)$$

where the elements $\bar{x}_i$ satisfy $\langle \bar{x}_i, x_j \rangle = \delta_{ij}$.

We can write $z = \sum_{i=1}^n (\eta_i x_i) + d$, where $\eta_i \in Z_2$ and d is a decomposable element.

This implies that $y = \sum_{i=1}^n (\eta_i x_i^{2^k}) + d^{2^k}$. If $x_i^{2^k} = 0$, z may be chosen such that $\eta_i = 0$.

Therefore $\eta_j \neq 0$ implies $x_j^{2^k} \neq 0$. Since $y^2 = 0$, if $\eta_i \neq 0$, then the height of x_i is 2^{k+1}.

Suppose now that $\eta_j \neq 0$. Then $\langle \gamma_{2^k}(\bar{x}_j), y \rangle \neq 0$; and hence $(\gamma_{2^k}(\bar{x}_j))^2 \neq 0$. If $\bar{x}_j^2 = 0$, then

$$\overline{\Delta}(\gamma_2(\bar{x}_j))^2 = \bar{x}_j^2 \otimes \bar{x}_j^2 = 0. \tag{4.50}$$

But since the degree of $(\gamma_2(\bar{x}_j))^2$ is either 16 or 32, $(\gamma_2(\bar{x}_j))^2 = 0$. This process may be

continued to show that $(\gamma_m(\bar{x}_j))^2 = 0$ for all $m \leq 2^k$, which would lead to a contradiction.

We conclude that $\bar{x}_j^2 \neq 0$. This implies that the degree of z must be 4, since otherwise

$\bar{x}_j^2 \in PH_{16} = 0$. Now for degree reasons $d = 0$.

We can now assume that the Borel generators x_i are all of degree 4 and A may have

some generators of degree 8. Assume also that the x_i are reordered to satisfy $r_1 \leq r_2 \leq$

$\cdots \leq r_n$. Let q be the smallest value such that $r_q \geq k$. If the η_i are reordered with

the x_i, then $z = \sum_{i=1}^{n} \eta_i x_i$ and $y = \sum_{i=1}^{n} \eta_i x_i^{2^k}$. Suppose $\bar{z}$ is an element which satisfies

$\bar{z} = \sum_{i=1}^{n} \beta_i \bar{x}_i$, where $\beta_i \in Z_2$ and $\beta_i = 0$ if $i < q$. Let p be the smallest number such that

$\beta_p = \eta_p = 1$. Note that $q \leq p$.

We now wish to choose a new set of Borel generators for H^*. These will be denoted

by y_i. This set will be chosen with the goal of making $\bar{z}$ a cogenerator of H_*. This will

permit us to show that $\bar{z}^2 \neq 0$.

Let

$$\begin{cases} y_i = x_i & \text{if } i \leq p \text{ or if } \beta_i = 0; \\ y_i = x_i + x_p & \text{otherwise.} \end{cases} \tag{4.51}$$

Since y_i are all linearly independent and the height of x_p is less than or equal to the height of x_i when $p < i$, we have an isomorphism of algebras

$$H^* \cong \frac{Z_2[y_1]}{(\varepsilon_1 y_1^{2^{r_1}})} \otimes \cdots \otimes \frac{Z_2[y_n]}{(\varepsilon_n y_n^{2^{r_n}})} \otimes A, \qquad (4.52)$$

where the y_i have replaced the x_i. Furthermore, $z = \sum_{i=1}^n \eta_i y_i$. This follows from the fact

that $\sum_{i=1}^n \beta_i \eta_i = \beta_p \eta_p$.

Now note that $\langle \bar{z}, y_i \rangle = \delta_{pi}$. This means that $\bar{z} = \bar{y}_p$ in the new coalgebra decompo-

sition for H_*. We may apply the same argument as in the case of the $\bar{x}_i$ elements to show

$\bar{z}^2 = \bar{y}_p^2 \neq 0$.

The element $\bar{z}$ was chosen to be an arbitrary linear combination of the elements x_i,

$q \leq i \leq n$. Since $\bar{z}^2 \neq 0$ in all cases, the elements $\bar{x}_i^2$ are linearly independent for $q \leq i \leq n$.

Because of this situation, we claim that there is an element w in cohomology such that

$$\overline{\Delta}(w) = (z + \sum_{i=1}^{q-1} \alpha_i x_i) \otimes (z + \sum_{i=1}^{q-1} \alpha_i x_i), \qquad (4.53)$$

where $\alpha_i \in Z_2$. The existence of such an element is shown as follows. Since the $\bar{x}_i^2$ are

linearly independent for $q \leq i \leq n$, it is not difficult to find an indecomposable element w

satisfying

$$\overline{\Delta}(w) = (\sum_{i=1}^n \alpha_i(x_i \otimes x_i)) + \mathrm{im}\,(1 + T^*), \qquad (4.54)$$

where $\alpha_i = \eta_i$ for $q \leq i \leq n$. Since all the terms in $\mathrm{im}\,(1 + T^*)$ must lie in $PH^* \otimes PH^*$, we

can alter w if necessary by a decomposable so that the $\mathrm{im}\,(1 + T^*)$ terms in the coproduct of w can be chosen to be whatever we want. In particular, w may be chosen so that

$$\overline{\Delta}(w) = (\sum_{i=1}^{n} \alpha_i(x_i \otimes x_i)) + (\sum_{1 \leq i < j \leq n} \alpha_i \alpha_j(x_i \otimes x_j + x_j \otimes x_i)). \tag{4.55}$$

Then

$$\begin{aligned}
\overline{\Delta}(w) &= (\sum_{i=1}^{n} \alpha_i x_i) \otimes (\sum_{i=1}^{n} \alpha_i x_i) \\
&= (\sum_{i=q}^{n} \eta_i x_i + \sum_{i=1}^{q-1} \alpha_i x_i) \otimes (\sum_{i=q}^{n} \eta_i x_i + \sum_{i=1}^{q-1} \alpha_i x_i) \\
&= (z + \sum_{i=1}^{q-1} \alpha_i x_i) \otimes (z + \sum_{i=1}^{q-1} \alpha_i x_i).
\end{aligned} \tag{4.56}$$

Set $a = \sum_{i=1}^{q-1} \alpha_i x_i$.

Now Lemma 4.6 implies that $Sq^1(z + a) = 0$. Then Lemma 4.7 may be applied to

show that $z^4 + a^4 = (z + a)^4 = 0$. By the choice of q, the height of a is strictly less

than the height of z. This means that the height of z must be less than or equal to 4. In

particular, $z^4 = 0$. Since $r - k = 2$ and $r \geq 4$, $y = z^{2^k} = z^{2^{r-2}} = (z^4)^{2^{r-4}} = 0$, which is a

contradiction. QED

PROPOSITION 4.9.   *For all $r \geq 4$, $QH^{2^r-1} = 0$.*

PROOF: Suppose not and choose the largest r such that $QH^{2^r-1} \neq 0$. Pick a non-zero

$x \in PH^{2^r-1}$. There is a factorization

$$Sq^{2^r+1} = Sq^2 Sq^1 Sq^{2^r-2} + Sq^{2^r} Sq^1. \tag{4.57}$$

Consider the two stage Postnikov system

$$K(Z_2, 2^{r+1} - 4, 2^r - 1)$$

$$\downarrow j$$

$$\tilde{f} \qquad E \xrightarrow{v} K(Z_2, 2^{r+1} - 1) \qquad\qquad (4.58)$$

$$\downarrow p$$

$$X \xrightarrow{f} K(Z_2, 2^r - 1) \xrightarrow{g} K(Z_2, 2^{r+1} - 3, 2^r)$$

where f represents x and $g^* = (\, Sq^{2^r-2} \quad Sq^1\,)$. The map v comes from the relation

(4.57) above and corresponds to a secondary operation ϕ. The class v is primitive and

satisfies $h_1(v) = p^* \iota_{2^r-1} \otimes p^* \iota_{2^r-1}$ and $j^*(v) = Sq^2 Sq^1 \iota_{2^{r+1}-4}$. $Sq^{2^r-2} x \in PH^{2^{r+1}-3} = 0$.

$Sq^1 x = 0$ by Lemma 4.8. Thus the lifting $\tilde{f}$ exists.

Because x is primitive, f is an H-map. If $\tilde{f}$ were an H-map, then $h_1(\phi(x))$ could

be computed. Unfortunately, this is not necessarily the case; but it will be shown that

Ωf is an h_1-map so that $h_2(\Omega\phi(x))$ can be computed. Since D_f, the H-deviation of f, is

homotopic to $pD_{\tilde{f}}$, $D_{\tilde{f}}$ factors as a composition

$$X \wedge X \xrightarrow{D} K(Z_2, 2^{r+1} - 4, 2^r - 1) \xrightarrow{j} E. \qquad\qquad (4.59)$$

The matrix $[D]$ satisfies the relation $(\Omega g) h_1(f) = (1 + T^*)[D]$, where $h_1(f)$ is the h_1-

deviation of f and Ωg is given by the matrix

$$\Omega g = \begin{pmatrix} Sq^{2^r-2} \\ Sq^1 \end{pmatrix}. \qquad\qquad (4.60)$$

Thus

$$(1 + T^*)[D] = \begin{pmatrix} Sq^{2^r-2}h_1(f) \\ \\ Sq^1h_1(f) \end{pmatrix}. \tag{4.61}$$

The indeterminacy of $h_1(f)$ depends on the choice of H-structure of f. The value of $h_1(f)$ may be altered by any element contained in $(1 + T^*)H^{2^r-2}(X \wedge X)$ by the appropriate change in H-structure for f. Thus the H-structure of f may be fixed so that $h_1(f)$ is possibly non-zero in

$$\begin{cases} PH^{2^{r-1}-1} \otimes PH^{2^{r-1}-1} + H^* \otimes DH^* + DH^* \otimes H^* & \text{if } r > 4; \\ \\ PH^7 \otimes PH^7 + PH^5 \otimes PH^9 + H^* \otimes DH^* + DH^* \otimes H^* & \text{if } r = 4. \end{cases} \tag{4.62}$$

This is due to the fact that the only remaining indecomposables are in degrees 3, 4, 5, 7, 8, 9 and degrees which are a power of two minus one. A routine calculation shows that

$$(1 + T^*)[D] \in H^* \otimes DH^* + DH^* \otimes H^*. \tag{4.63}$$

Therefore $[D] \in H^* \otimes DH^* + DH^* \otimes H^* + \ker(1 + T^*)$. The matrix $[D]$ has components in total degree equal to $2^{r+1} - 4$ and $2^r - 1$. Since $QH^{2^r-2} = QH^{d(1, \sum_{j=0}^{r-3} 2^j)} = 0$ by Proposition 2.6, the only diagonal terms of $[D]$ must lie in $H^* \otimes DH^* + DH^* \otimes H^*$. The argument from Proposition 3.5 may be applied, and the lifting $\tilde{f}$ may be altered so that $[D] \in H^* \otimes DH^* + DH^* \otimes H^*$. By looping the operation ϕ, the lifting $\Omega\tilde{f}$ is an h_1-map, and $h_2(\Omega\phi(x))$ may be computed.

The element $\sigma^* x$ is a non-zero primitive of degree $2^r - 2$. Therefore it is either indecomposable or it is the square of an odd degree indecomposable class. Suppose $\sigma^* x =$

a^2; then since $\deg a$ is odd $a = w + d$, where w is primitive and d is decomposable. Since

the degree of w is odd, $w \in \operatorname{im} \sigma^*$. If $w = \sigma^* z$, then $w^2 = Sq^{2^{r-1}-1} w = Sq^{2^{r-1}-1} \sigma^* z =$

$\sigma^*(Sq^1 Sq^{2^{r-1}-2} z)$. But $Sq^{2^{r-1}-2} z$ must be decomposable since it has degree $2^r - 2$. Thus

$w^2 = 0$ and hence $a^2 = d^2$, which is impossible. Therefore $\sigma^* x$ must be indecomposable.

Lemma 3.4 may be applied so that there is an $\mathcal{A}(2)$ sub Hopf algebra B_m of $H^*(\Omega X)$ such

that $h_2(\Omega f) \in B_m \otimes B_m$ and there is a homology primitive t such that $\langle t, \sigma^* x \rangle \neq 0$ and

$\langle t, B_m \rangle = 0$.

Then

$$h_2(\Omega \phi(x)) = \sigma^* x \otimes \sigma^* x + \operatorname{im}(1 + T^*) + \operatorname{im} Sq^2 Sq^1$$

$$(4.64)$$

$$+ H^*(\Omega X) \otimes B_m + B_m \otimes H^*(\Omega X),$$

and hence

$$\langle t \otimes t, h_2(\Omega \phi(x)) \rangle = \langle t \otimes t, \sigma^* x \otimes \sigma^* x + \operatorname{im}(1 + T^*) + \operatorname{im} Sq^2 Sq^1$$

$$+ H^*(\Omega X) \otimes B_m + B_m \otimes H^*(\Omega X) \rangle \qquad (4.65)$$

$$= \langle t \otimes t, \sigma^* x \otimes \sigma^* x + \operatorname{im} Sq^2 Sq^1 \rangle.$$

Note the following relations

$$tSq^1 = 0, \qquad tSq^2 Sq^1 = 0. \qquad (4.66)$$

$tSq^1 \in PH_{2^r - 3}(\Omega X)$; thus tSq^1 is dual to a suspension element. But $QH^{2^r - 2} = 0$ by

Proposition 2.6, so $tSq^1 = 0$. A similar argument applies to $tSq^2 Sq^1$ since $r \geq 4$.

The value of $(t \otimes t)Sq^2Sq^1$ is then equal to

$$(tSq^2 \otimes t + t \otimes tSq^2)Sq^1 = tSq^2Sq^1 \otimes t + t \otimes tSq^2Sq^1$$

$$(4.67)$$

$$= 0.$$

Therefore

$$\langle t \otimes t, h_2(\Omega\phi(x)) \rangle = \langle t \otimes t, \sigma^*x \otimes \sigma^*x \rangle \neq 0. \tag{4.68}$$

This implies in particular that $\Omega\phi(x) \neq 0$, which means σ^* is non-zero on some inde-

composable of H^* of degree $2^{r+1} - 1$. But by the choice of r, $QH^{2^{r+1}-1} = 0$. This is a

contradiction. QED

§5 QH^* in low degrees

This section concludes the proof of the Main Theorem, with the proofs that there are no indecomposables in degrees 3, 4, 5, 7, 8 or 9. This is the content of Propositions 5.4, 5.5, 5.6 and 5.7. It should be noted here that the bulk of the proof of Proposition 5.7 does not appear in this paper, and is due to Lin and Williams [11].

This section begins with Lemma 5.1, which states that certain primitives in degree 5 and 9 are in the image of Sq^1. This is useful in two places. First, in order to prove $QH^7 = 0$, we need to show that $Sq^1 PH^7 = 0$. In order to prove this, we need to use the fact that $Sq^2 PH^7 \subset \operatorname{im} Sq^1$. The second place where this lemma is needed is in the proof of Proposition 5.7. Here we need to know that degree 4 and 5 generators are connected by Sq^1, i.e. $Sq^1 : QH^4 \to QH^5$ is an isomorphism.

Following Lemma 5.1 is Lemma 5.2, which states that $Sq^1 PH^7 = 0$. As stated earlier, this is a necessary step in the proof that $QH^7 = 0$. Lemma 5.3 states that $Sq^1 PH^3 = 0$. This is also needed in the proof that $QH^7 = 0$, and in the proof that $QH^3 = 0$. Propositions 5.4 and 5.5 then give $QH^7 = 0$ and $QH^3 = 0$. Finally, Propositions 5.6 and 5.7 show that

$QH^n = 0$ for $n = 8$, 9 and $n = 4$, 5, respectively.

LEMMA 5.1.   *If x is either a primitive of degree 5 in the kernel of Sq^4, or a primitive of degree 9, then $x \in \operatorname{im} Sq^1$.*

PROOF: Suppose the degree of x is 9; the proof in the case that the degree of x is 5 is similar. There is a factorization

$$Sq^{10} = Sq^2 Sq^8 + Sq^9 Sq^1. \tag{5.1}$$

Consider the Postnikov system

$$
\begin{array}{ccccc}
& K(Z_2, 9, 16) & & & \\
& \downarrow j & & & \\
\bar{f} & E & \xrightarrow{\;v\;} & K(Z_2, 18) & \\
\nearrow & & \downarrow p & & \\
X & \xrightarrow{\;f\;} K(Z_2, 9) & \xrightarrow{\;g\;} & K(Z_2, 10, 17) &
\end{array}
\tag{5.2}
$$

where $f^* \iota_9 = x$, $g^* = (\; Sq^1 \quad Sq^8 \;)$, v represents an operation ϕ and arises from the relation

(5.1) above. The class v has reduced coproduct $p^* \iota_9 \otimes p^* \iota_9$ and $j^* v = Sq^9 \iota_9 + Sq^2 \iota_{16}$.

$Sq^1 x = 0$ by Lemma 3.2 and $Sq^8 = 0$ by Lemma 4.4. Thus the lifting $\tilde{f}$ exists.

Since f is an H-map, the H-deviation of $\tilde{f}$ factors as a composition

$$X \wedge X \xrightarrow{D} K(Z_2, 9, 16) \xrightarrow{j} E, \tag{5.3}$$

where the class $[D]$ satisfies

$$(1 + T^*)[D] = (\Omega g)h_1(f) = 0. \tag{5.4}$$

$h_1(f) = 0$ by Lemma 3.10. This implies $[D] \in \ker(1 + T^*)$. The operation ϕ satisfies the

formula

$$\overline{\Delta}\phi(x) = x \otimes x + (\, Sq^9 \quad Sq^2\,)[D]. \tag{5.5}$$

Assume $x \notin \operatorname{im} Sq^1$ modulo decomposables. Then there is an element $t \in PH_*$ such that

$\langle t, x \rangle \neq 0$ and $t Sq^1 = 0$. Thus

$$\langle t^2, \phi(x) \rangle = \langle t \otimes t, \overline{\Delta}\phi(x) \rangle$$

$$\tag{5.6}$$

$$= \langle t \otimes t, x \otimes x + (\, Sq^9 \quad Sq^2\,)[D] \rangle.$$

The value of $\langle t \otimes t, (\, Sq^9 \quad Sq^2\,)[D] \rangle$ is zero since $(\, Sq^9 \quad Sq^2\,)[D]$ is contained in $\operatorname{im}(1 +$

$T^*) + \operatorname{im}(Sq^1 \otimes Sq^1)$. Therefore $\langle t^2, \phi(x) \rangle = \langle t \otimes t, x \otimes x \rangle \neq 0$; in particular, $t^2 \neq 0$.

This is a contradiction since $QH^{18} = 0$ by Proposition 2.6. Therefore $x \in \operatorname{im} Sq^1$ modulo

decomposables.

Suppose $x = Sq^1 z + d_1$, where z is an indecomposable of degree 8, and $d_1 \in DH^9$.

Then

$$\overline{\Delta}(z) = \sum_i (w_i \otimes w_i) + \sum_j (u_j \otimes v_j + v_j \otimes u_j), \tag{5.7}$$

where $w_i \in PH^4$ and u_j, v_j are primitives of degree 3, 4 or 5. Let $y = z + \sum_j u_j v_j + \sum_{k<l} w_k w_l$; then

$$\overline{\Delta}(y) = \left(\sum_i w_i\right) \otimes \left(\sum_i w_i\right). \tag{5.8}$$

Then Lemma 4.6 may be applied to show that $Sq^1(\sum_i w_i) = 0$. This means that $0 = Sq^1\overline{\Delta}(y) = \overline{\Delta}(Sq^1 y)$. Let $d_2 = \sum_j u_j v_j + \sum_{k<l} w_k w_l$. Then

$$x = Sq^1(y + d_2) + d_1 = Sq^1 y + Sq^1 d_2 + d_1, \tag{5.9}$$

and $x + Sq^1 y \in PH^9$, which implies $Sq^1 d_2 + d_1 \in PH^9$. But $Sq^1 d_2 + d_1$ is decomposable and of odd degree, so $Sq^1 d_2 + d_1 = 0$. Hence $x = Sq^1 y$. QED

LEMMA 5.2. $Sq^1 PH^7 = 0$.

PROOF: Let $x \in PH^7$ and suppose $Sq^1 x \neq 0$. Then by Lemma 4.1, $Sq^1 x = y^2$ for some $y \in PH^4$. The proof of Lemma 2.2 shows that $y^4 = Sq^8 Sq^1 x = Sq^2 Sq^1 Sq^6 x = 0$ and that if $\bar{z} \in H_*$ satisfies $\langle \bar{z}, y^2 \rangle \neq 0$, then $\bar{z}^2 \neq 0$. It may be argued as in the proof of Lemma 4.8 that $(y + \sum_i y_i) \otimes (y + \sum_i y_i) \in \text{im } \overline{\Delta}$, where the y_i all have square zero, and hence by Lemma 4.6, $Sq^1(y + \sum_i y_i) = 0$. If $z = y + \sum_i y_i$, then $Sq^1 x = z^2$ and $Sq^1 z = 0$. There are two cases in the proof, either $Sq^2 x = 0$, or $Sq^2 x \neq 0$.

CASE 1: $Sq^2 x = 0$. Consider the Postnikov system

$$K(Z_2, 4, 7, 8)$$

$$\downarrow j_1$$

$$\tilde{f}_1 \qquad E_1 \xrightarrow{\ v_1\ } K(Z_2, 8) \qquad\qquad (5.10)$$

$$\downarrow p_1$$

$$X \xrightarrow{\ f_1\ } K(Z_2, 4, 7) \xrightarrow{\ g_1\ } K(Z_2, 5, 8, 9)$$

where $f_1^* = (\, z \quad x \,)$, $g_1^* = \begin{pmatrix} Sq^1 & 0 \\ Sq^4 & Sq^1 \\ 0 & Sq^2 \end{pmatrix}$, v_1 represents an operation ϕ_1 and arises from

the matrix relation

$$[Sq^1][Sq^4 \quad Sq^1] = [Sq^5 \quad 0]. \qquad\qquad (5.11)$$

The class v_1 has reduced coproduct $p_1^* \iota_4 \otimes p_1^* \iota_4$ and $j_1^* v_1 = Sq^1 \iota_7$. The lifting $\tilde{f}_1$ exists due

to the relations

$$Sq^1 z = 0,$$

$$Sq^4 z + Sq^1 x = 0, \qquad\qquad (5.12)$$

$$Sq^2 x = 0.$$

The two variable operation ϕ_1 then satisfies the formula

$$\overline{\Delta}\phi_1(z, x) = z \otimes z + \operatorname{im} Sq^1. \qquad\qquad (5.13)$$

Since $Sq^4 z \neq 0$, $z \notin \operatorname{im} Sq^1$. Then there is an element $t \in PH_4$ such that $\langle t, z \rangle \neq 0$ and

$tSq^1 = 0$. This implies $\langle t^2, \phi_1(z, x) \rangle \neq 0$, which means $\phi_1(z, x)$ is a non-primitive generator

of degree 8.

Now consider the element $Sq^1 v_1 \in H^*(E_1)$. It is a primitive element since $Sq^1 p_1^* \iota_4 = p_1^*(Sq^1 \iota_4) = 0$. Also $j_1^*(Sq^1 v_1) = Sq^1 Sq^1 \iota_7 = 0$. There are no primitives of $H^*(K(Z_2, 4, 7))$ of degree 9 which survive under p_1^*. Therefore it may be concluded that $Sq^1 v_1 = 0$, which implies $Sq^1 \phi_1(z, x) = 0$. But this is a contradiction to Lemma 4.2. This completes the proof of CASE 1.

CASE 2: $Sq^2 x \neq 0$. Let $Sq^2 x = w$. Then $w \in PH^9$. Lemma 5.1 implies $w = Sq^1 a$ for some indecomposable $a \in H^8$. Consider the Postnikov system

$$K(Z_2, 8, 12, 14)$$

$$\downarrow j_2$$

$$\tilde{f_2} \qquad E_2 \xrightarrow{v_2} K(Z_2, 16) \tag{5.14}$$

$$\downarrow p_2$$

$$X \xrightarrow{f_2} K(Z_2, 7, 8) \xrightarrow{g_2} K(Z_2, 9, 13, 15)$$

where $f_2^* = (\,x \quad a\,)$, $g_2^* = \begin{pmatrix} Sq^2 & Sq^1 \\ Sq^6 & 0 \\ 0 & Sq^7 \end{pmatrix}$, v_2 represents an operation ϕ_2 and arises from the matrix relation

$$[\, Sq^8 \quad Sq^4 \quad Sq^2 \,] \begin{bmatrix} Sq^2 & Sq^1 \\ Sq^6 & 0 \\ 0 & Sq^7 \end{bmatrix} = [\, Sq^{10} \quad Sq^9 \,]. \tag{5.15}$$

The class v_2 has reduced coproduct $p_2^* \iota_8 \otimes p_2^* \iota_8$ and $j_2^* v_2 = Sq^8 \iota_8 + Sq^4 \iota_{12} + Sq^2 \iota_{14}$. The

lifting $\tilde{f}_2$ exists due to the relations

$$Sq^2 x + Sq^1 a = 0,$$

$$Sq^6 x = 0, \tag{5.16}$$

$$Sq^7 a = 0.$$

The last two relations follow since $Sq^6 x \in PH^{13} = 0$ and $\overline{\Delta}(Sq^7 a) \in Sq^7(PH^3 \otimes PH^5 +$

$PH^4 \otimes PH^4 + PH^5 \otimes PH^3) = 0$ so that $Sq^7 a \in PH^{15} = 0$.

Let B be the $\mathcal{A}(2)$ sub-Hopf algebra of H^* generated by elements of degree less than

or equal to 5. Since the H-deviation of f_2^* is contained in $B \otimes B$, the two variable operation

ϕ_2 satisfies the formula

$$\overline{\Delta}\phi_2(x, a) = a \otimes a + \mathrm{im}\,(Sq^8 + Sq^4 + Sq^2) + H^* \otimes B + B \otimes H^*. \tag{5.17}$$

Since the degree of a is 8 and a is indecomposable, a must be a non-primitive generator.

This means that $a \notin \mathrm{im}\,\alpha$ modulo decomposables for any non-trivial $\alpha \in \mathcal{A}(2)$. Pick

$t \in PH_8$ such that $\langle t, a \rangle \neq 0$, $t\alpha = 0$ and $\langle t, B \rangle = 0$. Then $\langle t^2, \phi_2(x, a) \rangle = \langle t \otimes t, a \otimes a \rangle \neq 0$;

in particular $t^2 \neq 0$. But $QH^{16} = 0$, so this is a contradiction. QED

LEMMA 5.3. $Sq^1 PH^3 = 0$.

PROOF: Suppose $y \in PH^3$ and $Sq^1 y = x \neq 0$. Then $x \in PH^4$ and $Sq^4 x = 0$, $Sq^2 x = 0$

and $Sq^1 x = 0$. Consider the two stage system

$$K(Z_2, 4, 5, 7)$$

$$\downarrow j$$

$$\tilde{f} \quad E \xrightarrow{v} K(Z_2, 8) \tag{5.18}$$

$$\downarrow p$$

$$X \xrightarrow{f} K(Z_2, 4) \xrightarrow{g} K(Z_2, 5, 6, 8)$$

where $f^* \iota_4 = x$, $g^* = (\,Sq^1 \quad Sq^2 \quad Sq^4\,)$, and v represents an operation ϕ arising from the

relation $Sq^5 = Sq^1 Sq^4$. The class v has reduced coproduct $p^* \iota_4 \otimes p^* \iota_4$ and $j^* v = Sq^1 \iota_7$.

The lifting $\tilde{f}$ exists because $Sq^1 x = 0$, $Sq^2 x = 0$ and $Sq^4 x = 0$.

Since f is an H-map, the H-deviation of $\tilde{f}$ factors as a composition

$$X \wedge X \xrightarrow{D} K(Z_2, 4, 5, 7) \xrightarrow{j} E, \tag{5.19}$$

where the class $[D]$ satisfies

$$(1 + T^*)[D] = (\Omega g) h_1(f) = 0. \tag{5.20}$$

$h_1(f) = 0$ for degree reasons. This implies $[D] \in \ker(1 + T^*)$. Since $Sq^1 \ker(1 + T^*) \subset$

$\mathrm{im}\,(1 + T^*)$, the operation ϕ satisfies the formula

$$\overline{\Delta} \phi(x) = x \otimes x + \mathrm{im}\,(1 + T^*). \tag{5.21}$$

Choose $t \in PH_4$ such that $\langle t, x \rangle \neq 0$. Then

$$\langle t^2, \phi(x) \rangle = \langle t \otimes t, \overline{\Delta}\phi(x) \rangle$$

$$= \langle t \otimes t, x \otimes x + \operatorname{im}(1 + T^*) \rangle \tag{5.22}$$

$$= \langle t \otimes t, x \otimes x \rangle$$

$$\neq 0.$$

This implies that $\phi(x)$ is an indecomposable of degree 8.

Consider the element $Sq^1 v \in H^9(E)$. $Sq^1 v$ satisfies

$$\overline{\Delta}(Sq^1 v) = Sq^1 \overline{\Delta}(v)$$

$$= Sq^1(p^* \iota_4 \otimes p^* \iota_4) \tag{5.23}$$

$$= p^*(Sq^1 \iota_4) \otimes p^* \iota_4 + p^* \iota_4 \otimes p^*(Sq^1 \iota_4)$$

$$= 0.$$

Hence $Sq^1 v \in PH^9(E)$ and $j^*(Sq^1 v) = Sq^1 Sq^1 \iota_7 = 0$. Since $p^*(PH^9(K(Z_2, 4))) = 0$,

$Sq^1 v = 0$. This means that $Sq^1 \phi(x) = 0$. But since $\phi(x)$ is a degree 8 indecomposable,

this is a contradiction to Lemma 4.2. QED

PROPOSITION 5.4. $QH^7 = 0$.

PROOF: Suppose the proposition is false and choose a non-zero element $x \in PH^7$. There

are two cases to the proof, either $x \notin \operatorname{im} Sq^2 Sq^1$ or $x \in \operatorname{im} Sq^2 Sq^1$.

CASE 1: $x \notin \operatorname{im} Sq^2 Sq^1$. A simple argument shows $x \notin \operatorname{im} Sq^2 Sq^1$ modulo decompos-

ables. Consider the Postnikov system

$$K(Z_2, 7, 12)$$

$$\downarrow j_1$$

$$E_1 \xrightarrow{v_1} K(Z_2, 15) \qquad\qquad (5.24)$$

$$\tilde{f}_1 \nearrow \qquad \downarrow p_1$$

$$X \xrightarrow{f_1} K(Z_2, 7) \xrightarrow{g_1} K(Z_2, 8, 13)$$

where $f_1^* \iota_7 = x$, $g_1^* = (\, Sq^1 \quad Sq^6 \,)$ and v_1 represents an operation ϕ_1 arising from the

relation

$$Sq^9 = Sq^2 Sq^1 Sq^6 + Sq^8 Sq^1. \qquad\qquad (5.25)$$

The class v_1 is primitive and $h_1(v_1) = p_1^* \iota_7 \otimes p_1^* \iota_7$ and $j_1^* v_1 = Sq^2 Sq^1 \iota_{12}$. $Sq^1 x = 0$ by

Lemma 5.2 and $Sq^6 x \in PH^{13} = 0$. Therefore the lifting $\tilde{f}_1$ exists.

Since f_1 is an H-map, the H-deviation of $\tilde{f}_1$ factors as a composition

$$X \wedge X \xrightarrow{D_1} K(Z_2, 7, 12) \xrightarrow{j_1} E_1, \qquad\qquad (5.26)$$

where $[D_1]$ satisfies

$$(1 + T^*)[D_1] = (\Omega g_1) h_1(f_1). \qquad\qquad (5.27)$$

For degree reasons $h_1(f_1) \in PH^3 \otimes PH^3$. Since $Sq^1 PH^3 = 0$ by Lemma 5.3 and $Sq^3 PH^3 =$

0 by Lemma 3.2, $(\Omega g_1) h_1(f_1) = 0$, and $[D_1] \in \ker(1 + T^*)$. Because $QH^6 = 0$, the technique

applied in the proof of Proposition 3.5 can be applied to show $[D_1] \in H^* \otimes DH^* + DH^* \otimes H^*$.

The operation can be looped now so that $\Omega \tilde{f}_1$ is an h_1-map.

If $\sigma^* x$ is decomposable, then $\sigma^* x$ is the square of a degree 3 indecomposable. This implies $\sigma^* x = Sq^3 \sigma^* y$ for some $y \in PH^4$. But $Sq^3 \sigma^* y = \sigma^* (Sq^1 Sq^2 y) = 0$. Therefore $\sigma^* x$ must be indecomposable. Lemma 3.4 implies the existence of an $\mathcal{A}(2)$ Hopf algebra B_m and an element $t \in PH_6(\Omega X)$ such that $\langle t, \sigma^* x \rangle \neq 0$ and $\langle t, B_m \rangle = 0$.

Suppose $\sigma^* x = Sq^2 Sq^1 w + d$, where $d \in DH^*(\Omega X)$. Then for degree reasons, $w \in PH^3(\Omega X)$, and hence $w = \sigma^* a$ for some $a \in PH^4$. This means d is primitive, so that $d = Sq^3 \sigma^* b = Sq^1 \sigma^* (Sq^2 b)$. But $Sq^2 b \in PH^6 = 0$, so $d = 0$. Then $\sigma^* (x + Sq^2 Sq^1 w) = 0$, which implies $x + Sq^2 Sq^1 w \in DH^*$. However, this is not the case. Therefore, t may be chosen so that $t Sq^2 Sq^1 = 0$.

The operation $\Omega \phi_1$ satisfies the formula

$$h_2(\Omega \phi_1(x)) = \sigma^* x \otimes \sigma^* x + \mathrm{im}\, Sq^2 Sq^1 + \mathrm{im}\, (1 + T^*) + H^*(\Omega X) \otimes B_m + B_m \otimes H^*(\Omega X). \quad (5.28)$$

The element $t Sq^1 \in PH_5(\Omega X)$. If $t Sq^1 \neq 0$, then the image of σ^* must be non-zero in degree 5. This would imply that $QH^6 \neq 0$, which is impossible. Therefore $t Sq^1 = 0$. The value of $(t \otimes t) Sq^2 Sq^1$ is

$$t Sq^2 Sq^1 \otimes t + t Sq^2 \otimes t Sq^1 + t Sq^1 \otimes t Sq^2 + t \otimes t Sq^2 Sq^1 = 0. \quad (5.29)$$

Then $\langle t \otimes t, h_2(\Omega\phi_1(x)) \rangle = \langle t \otimes t, \sigma^*x \otimes \sigma^*x \rangle \neq 0$; and in particular, $\Omega\phi_1(x) \neq 0$. The degree

of $[\Omega\phi_1(x)]$ is 14. This means σ^* is non-zero on QH^{15}. But $QH^{15} = 0$ by Proposition 4.9.

This completes the proof of CASE 1.

CASE 2: $x \in \operatorname{im} Sq^2Sq^1$. Suppose $x = Sq^2Sq^1z$, $z \in PH^4$. Consider the two stage

system

$$
\begin{array}{ccc}
K(Z_2,7) & & \\
\downarrow {\scriptstyle j_2} & & \\
E_2 & \xrightarrow{\ v_2\ } & K(Z_2,15) \\
{\scriptstyle \tilde{f}_2}\nearrow \quad \downarrow {\scriptstyle p_2} & & \\
X \xrightarrow{\ f_2\ } K(Z_2,4) & \xrightarrow{\ g_2\ } & K(Z_2,8)
\end{array}
\tag{5.30}
$$

where $f_2^*\iota_4 = z$, $g_2^* = Sq^3Sq^1$ and v_2 represents an operation ϕ_2 arising from the relation

$$
Sq^9Sq^2Sq^1 = (Sq^8 + Sq^4Sq^4)Sq^3Sq^1.
\tag{5.31}
$$

The class v_2 is primitive and $h_1(v_2) = Sq^2Sq^1p_2^*\iota_4 \otimes Sq^2Sq^1p_2^*\iota_4$ and $j_2^*v_2 = Sq^4Sq^4\iota_7$.

The lifting $\tilde{f}_2$ exists due to the fact that $Sq^3Sq^1z = Sq^1x = 0$ by Lemma 5.2.

Since f_2 is an H-map, the H-deviation of $\tilde{f}_2$ factors as a composition

$$
X \wedge X \xrightarrow{\ D_2\ } K(Z_2,7) \xrightarrow{\ j_2\ } E_2,
\tag{5.32}
$$

where $[D_2]$ satisfies

$$(1 + T^*)[D_2] = (\Omega g_2)h_1(f_2). \tag{5.33}$$

For degree reasons $h_1(f_2) = 0$. Therefore $[D_2] \in \ker(1 + T^*)$. The technique applied in

the proof of Proposition 3.5 can be applied to alter $[D_2]$ so that $[D_2] = 0$. This means

there is a choice of lifting $\tilde{f}_2$ which is an H-map. Then

$$h_1(\phi_2(z)) = x \otimes x + \mathrm{im}\, Sq^4 Sq^4 + \mathrm{im}\,(1 + T^*). \tag{5.34}$$

Pick $t \in PH_7$ such that $\langle t, x \rangle \neq 0$. $tSq^1 \in PH_6$; since $QH^6 = 0$, $tSq^1 = 0$. Then

$$(t \otimes t)Sq^4 Sq^4 = (tSq^2 \otimes tSq^2)Sq^4$$

$$= tSq^2 Sq^2 \otimes tSq^2 Sq^2$$

$$= tSq^1 Sq^2 Sq^1 \otimes tSq^1 Sq^2 Sq^1 \tag{5.35}$$

$$= 0.$$

Hence $\langle t \otimes t, \phi_2(z) \rangle = \langle t \otimes t, x \otimes x \rangle \neq 0$; and in particular, $\phi_2(z) \neq 0$. But $\phi_2(z) \in PH^{15} = 0$

by Proposition 4.9. This is a contradiction. QED

PROPOSITION 5.5. $QH^3 = 0$.

PROOF: Suppose the proposition is false and choose a non-zero element $x \in PH^3$. There

is a factorization

$$Sq^5 = Sq^2 Sq^3 + Sq^4 Sq^1. \tag{5.36}$$

Consider the diagram

$$K(Z_2, 3, 5)$$

$$\downarrow j$$

$$\overset{\tilde{f}}{} \quad E \xrightarrow{\ v\ } K(Z_2, 7) \tag{5.37}$$

$$\downarrow p$$

$$X \xrightarrow{\ f\ } K(Z_2, 3) \xrightarrow{\ g\ } K(Z_2, 4, 6)$$

where $f^* \iota_3 = x$, $g^* = (\, Sq^1 \quad Sq^3 \,)$ and v represents an operation ϕ arising from the relation

(5.36) above. The class v is primitive and $h_1(v) = p^* \iota_3 \otimes p^* \iota_3$ and $j^* v = Sq^2 \iota_5$. The lifting

$\tilde{f}$ exists because $Sq^1 x = 0$ by Lemma 5.3 and $Sq^3 x = x^2 = 0$ by Lemma 3.1.

Since f is an H-map, the H-deviation of $\tilde{f}$ factors as a composition

$$X \wedge X \xrightarrow{\ D\ } K(Z_2, 3, 5) \xrightarrow{\ j\ } E. \tag{5.38}$$

For degree reasons the class $[D] = 0$. Also for degree reasons $h_1(f) = 0$. Therefore

$$h_1(\phi(x)) = x \otimes x + \operatorname{im} Sq^2 + \operatorname{im}(1 + T^*). \tag{5.39}$$

Pick $t \in PH_3$ such that $\langle t, x \rangle \neq 0$. For degree reasons $t\alpha = 0$ for any non-trivial

$\alpha \in \mathcal{A}(2)$. Then $\langle t \otimes t, h_1(\phi(x)) \rangle \neq 0$; and in particular, $0 \neq \phi(x) \in PH^7$. But $PH^7 = 0$

by Proposition 5.4, a contradiction. QED

PROPOSITION 5.6. $QH^8 = 0$ and $QH^9 = 0$.

PROOF: It suffices to show $QH^9 = 0$ since, by Lemma 4.2, any 8 dimensional indecomposable is connected to a degree 9 indecomposable by Sq^1. Suppose now that $QH^9 \neq 0$ and choose a non-zero $y \in PH^9$. By Lemma 3.8, $y = Sq^4 x$ for some $x \in PH^5$. Consider the three stage Postnikov system

$$
\begin{array}{cccccc}
K(Z_2,8,9,12) & & & & & \\
& \searrow{}^{j_1} & & & & \\
K(Z_2,5,6) & & E_1 & \xrightarrow{w} & K(Z_2,18) & \\
& {}^{j_0}\searrow \quad \nearrow & \downarrow {}^{p_1} & & & \\
& {}^{f_1}\nearrow \searrow & E_0 & \xrightarrow{v} & K(Z_2,9,10,13) & \\
& {}^{f_0}\nearrow & \downarrow {}^{p_0} & & & \\
X & \xrightarrow{f} & K(Z_2,5) & \xrightarrow{g} & K(Z_2,6,7) &
\end{array}
\tag{5.40}
$$

where $f^* \iota_5 = x$ and $g^* = (\, Sq^1 \quad Sq^2 \,)$. The element $v = (v_i)$, where v_1, v_2 and v_3 correspond to operations $Sq^1\psi_2$, $Sq^2\psi_2$ and ψ_1, respectively. The operations ψ_1 and ψ_2 are given by the matrix relation

$$
\begin{array}{c} \psi_1 \\ \psi_2 \end{array}
\begin{bmatrix} Sq^8 + Sq^4 Sq^4 & Sq^4 Sq^2 Sq^1 + Sq^7 \\ Sq^3 & Sq^2 \end{bmatrix}
\begin{bmatrix} Sq^1 \\ Sq^2 \end{bmatrix} =
\begin{bmatrix} Sq^9 \\ 0 \end{bmatrix}.
\tag{5.41}
$$

The map v in this case is an h_1-map. The element w arises from the relation

$$
Sq^{10}Sq^4 = Sq^2 Sq^4 \psi_1 + Sq^2 Sq^8 Sq^1 \psi_2 + Sq^2 Sq^7 Sq^2 \psi_2 +
$$

$$
\tag{5.42}
$$

$$
Sq^9 Sq^2 Sq^1 Sq^2 + Sq^9 Sq^4 Sq^1,
$$

which is a result of the two relations

$$Sq^{10}Sq^4 = Sq^2Sq^8Sq^4 + Sq^9Sq^2Sq^1Sq^2 + Sq^9Sq^4Sq^1$$

$$\text{(5.43)}$$

$$Sq^8Sq^4 = Sq^4\psi_1 + (Sq^9 + Sq^7Sq^2)\psi_2 + Sq^6Sq^5\psi_3.$$

Thus $\overline{\Delta}(w) = Sq^4(p_0p_1)^*\iota_5 \otimes Sq^4(p_0p_1)^*\iota_5$ and $j_1^*w = Sq^2Sq^8\iota_8 + Sq^2Sq^7\iota_9 + Sq^2Sq^4\iota_{12}$.

Let ϕ be the tertiary operation associated to the element w.

The first lifting f_0 exists since $Sq^1x = 0$ by Lemma 3.2 and $Sq^2x \in PH^7 = 0$

by Proposition 5.4. Moreover, since f is an H-map, the H-deviation of f_0 factors as a

composition

$$X \wedge X \xrightarrow{D_0} K(Z_2, 5, 6) \xrightarrow{j_0} E_0. \qquad \text{(5.44)}$$

For degree reasons $[D_0] = 0$, so f_0 is an H-map. Because f is an h_1-map, a similar

argument shows that f_0 is also an h_1-map. Because v and f_0 are H-maps, the classes

$\psi_1(x)$ and $\psi_2(x)$ are primitives. Thus $\psi_1(x) \in PH^{13} = 0$ and $\psi_2(x) = z^2$ for some

$z \in PH^4$. $Sq^1\psi_2(x) = Sq^1z^2 = 0$ and $Sq^2\psi_2(x) = Sq^2z^2 = (Sq^1z)^2 = 0$ by Lemma 3.1;

hence there is a second lifting f_1. The H-deviation of f_1 factors as a composition

$$X \wedge X \xrightarrow{D_1} K(Z_2, 8, 9, 12) \xrightarrow{j_1} E_1, \qquad \text{(5.45)}$$

and $[D_1]$ satisfies $(1 + T^*)[D_1] = (\Omega v)h_1(f_0) = 0$. Then the operation ϕ satisfies

$$\overline{\Delta}\phi(x) = y \otimes y + (\, Sq^2Sq^8 \quad Sq^2Sq^7 \quad Sq^2Sq^4 \,)[D_1]$$

$$= y \otimes y + (\, Sq^2Sq^8 \quad Sq^2Sq^7 \quad Sq^2Sq^4 \,)\ker(1 + T^*) \qquad \text{(5.46)}$$

$$= y \otimes y + \text{im}\,(Sq^1Sq^4 \otimes Sq^1Sq^4 + Sq^1Sq^2 \otimes Sq^1Sq^2) + \text{im}\,(1 + T^*).$$

Choose $t \in PH_9$ such that $\langle t, y \rangle \neq 0$. Then for any $\alpha \in \mathcal{A}(2)$, $tSq^1\alpha = 0$. So

$$\langle t^2, \phi(x) \rangle = \langle t \otimes t, \overline{\Delta}\phi(x) \rangle$$

$$= \langle t \otimes t, y \otimes y + \mathrm{im}\,(Sq^1 Sq^4 \otimes Sq^1 Sq^4 + Sq^1 Sq^2 \otimes Sq^1 Sq^2) + \mathrm{im}\,(1 + T^*) \rangle$$

$$= \langle t \otimes t, y \otimes y \rangle$$

$$\neq 0;$$

(5.47)

in particular, $t^2 \neq 0$. But $t^2 \in PH_{18} = 0$, a contradiction. QED

PROPOSITION 5.7. $QH^4 = 0$ and $QH^5 = 0$.

PROOF: By Lemmas 4.2 and 5.1, $Sq^1 : QH^4 \to QH^5$ is an isomorphism. Lemma 3.1 implies that all 5 dimensional generators have height two. It will be shown that all 4 dimensional generators have infinite height.

Suppose $x \in PH^4$ is such that $x^{2^s} \neq 0$ and $x^{2^{s+1}} = 0$. Since $Sq^{2^{s+2}} x^{2^s} = x^{2^{s+1}} = 0$ there is a secondary operation ϕ associated to the factorization $Sq^{2^{s+2}+1} = Sq^1 Sq^{2^{s+2}}$ such that

$$\overline{\Delta}\phi(x^{2^s}) = x^{2^s} \otimes x^{2^s} + \mathrm{im}\, Sq^1.$$

(5.48)

Let Λ be the exterior algebra generated by the odd degree primitives of H^*. Since $Sq^1 PH^{odd} = 0$, Λ is invariant under the action of $\mathcal{A}(2)$; thus Λ is an $\mathcal{A}(2)$ invariant sub-Hopf algebra of H^* so that the following is an exact sequence of Hopf algebras over

$\mathcal{A}(2)$

$$H^* \to H^* /\!/ \Lambda \to 0 \tag{5.49}$$

Let Γ be the dual of $H^* /\!/ \Lambda$. Let $t \in \Gamma$ be an element such that $\langle t, \{x^{2^{\bullet}}\} \rangle \neq 0$. The Hopf

algebra $H^* /\!/ \Lambda$ is concentrated in even degrees. Therefore, the induced action of Sq^1 on it

is identically zero, and hence $tSq^1 = 0$. Then

$$\langle t^2, \{\phi(x^{2^{\bullet}})\} \rangle = \langle t \otimes t, \overline{\Delta}\{\phi(x^{2^{\bullet}})\} \rangle = \langle t \otimes t, \{x^{2^{\bullet}}\} \otimes \{x^{2^{\bullet}}\} \rangle \neq 0. \tag{5.50}$$

Since H^* is primitively generated, $H^* /\!/ \Lambda$ is primitively generated and Γ has no squares.

This gives a contradiction and the conclusion that x has infinite height.

It has been shown that H^* is isomorphic as Hopf algebras over the Steenrod algebra

to a tensor product of copies of

$$Z_2[x] \otimes \Lambda[y], \tag{5.51}$$

where $x \in PH^4$, $y \in PH^5$ and $Sq^1 x = y$. Lin and Williams [11] have shown that there

is no homotopy commutative mod 2 H-space with cohomology as such. The only possible

conclusion is that $QH^4 = 0$ and $QH^5 = 0$. QED

§6 Proof of corollaries

In this section Theorem 0.1 is used to prove Corollaries 0.2, 0.3 and 0.4. The main

tool in the proof of Corollary 0.2 is the Eilenberg-Moore spectral sequence. The reader

is assumed to be familiar with this spectral sequence. The background material for the

relevant calculations can be found in sections 11–14 of Kane [8]. The proof of Corollary

0.2 depends on Lemma 6.1, which goes as follows. Given a simply connected X, we pass

to the 2-connective cover $X\langle 2\rangle$. In order to prove the corollary, we need first to show that

$H^*(X\langle 2\rangle)$ is finitely generated as an algebra. This is done with the help of the EMSS.

Once this is known, we can apply the Main Theorem to show $H^*(X\langle 2\rangle)$ is acyclic. Some

further analysis with the EMSS gives the result that $X\langle 2\rangle$ has the mod 2 homotopy type

of a product of $K(Z,2)$s. This is the result of Lemma 6.1. With this result in hand, we

can prove the rest of Corollary 0.2 with a similar analysis using a covering space argument

together with the EMSS.

As an immediate consequence of Corollary 0.2 we get Corollary 0.3, which states that

if X is simply connected, its cohomology is two torsion free. This follows from the fact that $K(Z,2)$ has two torsion free cohomology.

The proof of Corollary 0.4 is nearly identical to part of the proof of Corollary 0.2.

In both proofs it is necessary to show that a connective cover has finitely generated co-

homology. The proof for the 3-connective cover of a simply connected finite loop space

corresponds to the proof for the 2-connective cover of a simply connected H-space with

finitely generated cohomology.

LEMMA 6.1.   *If X is a simply connected homotopy commutative homotopy associative*

H-space whose mod 2 cohomology is finitely generated as an algebra, then X has the mod

2 homotopy type of a product of spaces of the type $K(Z,2)$.

PROOF: Construct the 2-connective cover, $X\langle 2\rangle$, of X by use of the pullback diagram

below,

$$
\begin{array}{ccc}
X\langle 2\rangle & \rightarrow & LK \\
\downarrow{\scriptstyle p} & & \downarrow \\
X & \xrightarrow{\ f\ } & K
\end{array}
\tag{6.1}
$$

where f represents the 2-dimensional cohomology of X. In other words, K is a product

of Eilenberg-MacLane spaces of types $K(Z,2)$ or $K(Z_{2^r},2)$ for some $r \geq 1$, depending

on the 2-dimensional cohomology of X. Pick K and f so that $f^*(\beta_r \iota_2) \neq 0$ for each

$\beta_r \iota_2 \in H^*(K(Z_{2^r},2))$. Thus, if $x \in H^*(X)$, and x lifts to a Z_{2^r} class, it is represented

by a map $X \rightarrow K(Z_{2^r},2)$. If x lifts to an integral class, then it is represented by a map

$X \rightarrow K(Z,2)$.

The space $X\langle 2\rangle$ is a 2-connected homotopy commutative homotopy associative H-space. This follows from the fact that f is an h_1-map, and also an a_3-map because the obstructions are in such low degrees that they must be zero. By Theorem 0.1, the mod 2 cohomology of $X\langle 2\rangle$ must either be acyclic, or have infinitely many algebra generators.

By Kane [8], the E_∞ term of the Eilenberg-Moore spectral sequence converging to $H^*(X\langle 2\rangle)$ is isomorphic as Hopf algebras to

$$H^*(X)/\!/\mathrm{im}\, f^* \otimes \mathrm{Tor}_\Gamma(Z_2, Z_2), \qquad (6.2)$$

where Γ is a certain sub-algebra of $H^*(K)$. Since $H^*(K)$ is a polynomial algebra, so is Γ. This implies that $\mathrm{Tor}_\Gamma(Z_2, Z_2)$ is an exterior algebra on the generators of Γ. The construction of Γ is such that the ideal generated by Γ is $\ker f^*$. The construction is as follows. Since $\mathrm{im}\, f^*$ is a sub-Hopf algebra of $H^*(X)$, it has a Borel decomposition. Let $\mathrm{im}\, f^* = \otimes_i \Lambda_i$, where each Λ_i has a single generator λ_i. For each λ_i choose γ_i such that $f^*(\gamma_i) = \lambda_i$. The γ_i are polynomial generators of $H^*(K)$. Choose the remaining polynomial generators $\{\gamma_j\}$ of $H^*(K)$ from $\ker f^*$. If the height of $\lambda_i = 2^q$, let $\gamma_i' = \gamma_i^{2^q}$. Then Γ is the sub-algebra of $H^*(K)$ generated by $\{\gamma_i'\} \cup \{\gamma_j\}$.

Let Ω be the sub-algebra of $H^*(X\langle 2\rangle)$ generated by $H^*(X)/\!/\mathrm{im}\, f^*$ plus representatives in $H^*(X\langle 2\rangle)$ for the elements $s\gamma_i'$ in $\mathrm{Tor}_\Gamma(Z_2, Z_2)$. Let $\Phi = H^*(X\langle 2\rangle)/\!/\Omega$. Consider the diagram below.

$$
\begin{array}{ccc}
\Omega K & \xrightarrow{\ h\ } & X\langle 2\rangle \\
\downarrow & & \downarrow p \\
* & \rightarrow & X \\
\downarrow & & \downarrow f \\
K & \xrightarrow{\ id\ } & K
\end{array}
\tag{6.3}
$$

The diagram induces a map of spectral sequences, and Kane shows that $\Phi \cong \operatorname{im} h^*$ and

$\ker h^*$ is the ideal generated by Ω. He also shows that $H^*(X\langle 2\rangle) \cong \Omega \otimes \Phi$ as algebras.

It is desirable to show that Ω and Φ are finitely generated in order to show $H^*(X\langle 2\rangle)$ is

finitely generated as an algebra.

Consider $H^*(K)$. The mod 2 cohomology of $K(Z,2)$ is a polynomial algebra on the

fundamental class ι_2. The mod 2 cohomology of $K(Z_{2^r},2)$ is a polynomial algebra on

the classes ι_2, $\beta_r\iota_2$, $Sq^2\beta_r\iota_2$, $Sq^4Sq^2\beta_r\iota_2$, etc. These classes are primitives, and except

for the fundamental class, they all have odd degree. Therefore, their images under f^*

are generators of $H^*(X)$. This means that only finitely many are not in the kernel of

f^*, which implies that $\{\gamma_i'\}$ is a finite set. This implies that Ω is finitely generated since

$H^*(X)/\!/\operatorname{im} f^*$ must also be finitely generated.

For each copy of $K(Z_{2^r},2)$ pick the lowest degree generator in $\ker f^*$. Suppose in one

case this generator is $Sq^{2^s} \cdots Sq^4 Sq^2 \beta_r\iota_2$. Then all the higher degree generators from this

copy of $K(Z_{2^r},2)$ are also in $\ker f^*$. Thus these generators form a subset of $\{\gamma_j\}$. The

image under h^* of the representatives of these generators forms a polynomial sub-algebra

of $H^*(\Omega K)$. The degree of $s(Sq^{2^s} \cdots Sq^4 Sq^2 \beta_r\iota_2)$ is 2^{s+1}. If $h^*(s(Sq^{2^s} \cdots Sq^4 Sq^2 \beta_r\iota_2)) =$

a, then $h^*(s(Sq^{2^{r+1}}Sq^{2^r}\cdots Sq^4Sq^2\beta_r\iota_2)) = a^2$, $h^*(s(Sq^{2^{r+2}}Sq^{2^{r+1}}Sq^{2^r}\cdots Sq^4Sq^2\beta_r\iota_2)) =$

$\dot{a}^4$, and it is not difficult to see that a continuation of this argument will show that this

set of generators corresponds to a polynomial sub-algebra of $H^*(\tilde{X})$ on a single generator.

This means that Φ is finitely generated (since there are only finitely many 2-dimensional

generators of $H^*(X)$).

It has been shown that $H^*(X\langle 2\rangle)$ is finitely generated as an algebra. Therefore, by

the Main Theorem, $H^*(X\langle 2\rangle)$ is acyclic. The generators of Γ all have degree greater than

or equal to 4; thus they give rise to non-trivial classes in $H^*(X\langle 2\rangle)$. This implies that

$\mathrm{Tor}_\Gamma(Z_2, Z_2) = 0$, and $E_0(H^*(X\langle 2\rangle)) \cong H^*(X)/\!/\mathrm{im}\,f^*$. But $H^*(X)/\!/\mathrm{im}\,f^*$ is concentrated

in external degree 0, so there are no non-trivial extensions and $H^*(X\langle 2\rangle) \cong H^*(X)/\!/\mathrm{im}\,f^*$.

The only way this is possible is if f^* is an isomorphism, $H^*(X)$ has mod 2 cohomology

isomorphic to the mod 2 cohomology of a product of spaces of the type $K(Z,2)$, and f is

a map inducing this isomorphism. QED

COROLLARY 0.2.   *If X is a homotopy commutative, homotopy associative H-space whose*

mod 2 cohomology is finitely generated as an algebra, then X has the mod 2 homotopy

type of a product of spaces from the list ($r \geq 1$):

$$K(Z,1),\ K(Z_{2^r},1),\ K(Z,2). \tag{6.4}$$

PROOF: Consider the covering space, $\tilde{X}$, of X, which fits in a fibration sequence

$$
\begin{array}{ccc}
G & \to & \tilde{X} \\
& & \downarrow p \\
& & X \xrightarrow{\ f\ } K(G,1)
\end{array}
\tag{6.5}
$$

where G is the 2-component of $\pi_1(X)$ and is an abelian group. $\tilde{X}$ is a (mod 2) simply connected homotopy associative, homotopy commutative H-space. Dwyer [4] has shown that the EMSS can be applied to this fibration. As in the proof of Lemma 6.1, $\tilde{X}$ has finitely generated cohomology. We can apply Lemma 6.1 to $\tilde{X}$ to show $\tilde{X}$ has the mod 2 homotopy type of a product of $K(Z,2)$s.

We wish to show that f^* is a monomorphism. If it weren't, then there would be an even degree element in $\ker f^*$. This would produce an odd degree class in $H^*(\tilde{X})$. But $H^*(\tilde{X})$ is even dimensional, so f^* must be a monomorphism. We may conclude by the EMSS that $H^*(\tilde{X}) \cong H^*(X)//\operatorname{im} f^*$; and so there is a map $g : X \to K(G,1) \times K(Z \oplus \cdots \oplus Z, 2)$ which induces a mod 2 cohomology isomorphism. QED

COROLLARY 0.3. *If X is a simply connected homotopy commutative, homotopy associative H-space whose mod 2 cohomology is finitely generated as an algebra, then the integral cohomology of X is two torsion free.*

PROOF: This is immediate from Lemma 6.1, since the cohomology of $K(Z,2)$ is two torsion free. QED

Assume now that spaces are not localized. Suppose that $\{Y_i\}_{i \in S}$ is a set of simply

connected finite loop spaces. Let Q and R be finite subsets of the indexing set S such that

their union is non-empty. Set $W = \times_{j \in Q}(Y_j\langle 3\rangle) \times_{k \in R} (Y_k)$.

COROLLARY 0.4.   *There is no choice of multiplication on W which is homotopy commu-*

tative.

PROOF: A simply connected finite loop space is 2-connected. Also, the 3-connective cover

of a homotopy associative, homotopy commutative H-space is homotopy associative and

homotopy commutative. This follows since the obstructions to homotopy associativity

and homotopy commutativity take values in zero groups for dimensional reasons. Hence

it suffices to show by the Main Theorem that the cohomology of W is finitely generated.

We use the EMSS in the same way it was used in the proof of Lemma 6.1. The important

thing to notice is that π_3 of a finite loop space is a direct sum of copies of the integers Z.

We leave the details to the reader. QED

§7 Appendix

A^*	Hopf algebra dual of the Hopf algebra A
$\overline{A}$	augmentation ideal of the Hopf algebra A
$A \cdot A$	image of the multiplication map in the Hopf algebra A
$\mathcal{A}(2)$	mod 2 Steenrod algebra
$d(r,k)$	$2^r(2k+1)$
D_f	H-deviation of the map f
DH^*	module of decomposables of H^*
$\deg x$	degree of x
$h_n(f)$	h_n-deviation of the h_{n-1}-map f
H^*	mod 2 cohomology of X
LK	space of basepointed paths in K
P_2X	projective plane of X
PA	module of primitives of the Hopf algebra A
QA	module of indecomposables of the Hopf algebra A

$X\langle n \rangle$ n-connective cover of an $(n-1)$-connected space X

$\hat{x}$ projection of a cohomology element x in QH^*

$\bar{x}$ homology dual of a cohomology element x

$\{x\}$ equivalence class of $x \in A$ in quotient Hopf algebra $A /\!\!/ B$

β_r r^{th} mod 2 cohomology Bockstein homomorphism

δ_{ij} Kronecker delta

$\overline{\Delta}$ reduced coproduct

$\epsilon : \Sigma\Omega X \to X$ evaluation map

λA Hopf algebra dual of $\xi(A^*)$

ξA image of the squaring map in the Hopf algebra A

σ^* cohomology suspension homomorphism

$\sigma^*(S)$ image of S under σ^*

ΣX suspension of X

ΩX (Moore) loop space of X

$\langle\ ,\ \rangle$ Kronecker pairing

Bibliography

[1] W. Browder, *Homotopy commutative H-spaces*, Ann. Math. **75** (1962), 283–311.

[2] W. Browder and E. Thomas, *On the projective plane of an H-space*, Ill. J. Math. **7** (1963), 492–502.

[3] A. Clark, *On π_3 of finite dimensional H-spaces*, Ann. Math. **78** (1963), 193–196.

[4] W. Dwyer, *Strong convergence of the Eilenberg-Moore spectral sequence*, Topology **13** (1974), 255–265.

[5] Y. Hemmi, *Higher homotopy commutativity and the mod p torus theorem*, to appear.

[6] J.R. Hubbuck, *On homotopy commutative H-spaces*, Topology **8** (1969), 119–126.

[7] K. Iriye and A. Kono, *Mod p retracts of G-product spaces*, Math. Z. **190** (1985), 357–363.

[8] R. Kane, *Implications in Morava K-theory*, Mem. AMS **340** (1986).

[9] J.P. Lin, *Torsion in H-spaces, I*, Ann. Math. **103** (1976), 457–486.

[10] J.P. Lin, *A cohomological proof of the torus theorem*, Math. Z. **190** (1985), 469–476.

[11] J.P. Lin, and F. Williams, *On homotopy commutative H-spaces*, to appear in Proc.

AMS.

[12] J.P. Lin, and F. Williams, private communication.

[13] C. A. McGibbon, *Higher forms of homotopy commutativity and finite loop spaces*, Math. Z. **201** (1989), 363–374.

[14] J. Milnor, and J.C. Moore, *On the structure of Hopf algebras*, Ann. Math. **81** (1965), 211–264.

[15] M. Slack, *Maps between iterated loop spaces*, to appear in J. Pure and Applied Alg.

[16] E. Thomas, *Whitney–Cartan product formulae*, Math. Z. **118** (1970), 115–138.

[17] F. Williams, *The H-deviation of a lifting*, to appear in Proc. AMS.

Department of Mathematics

University of California, San Diego

La Jolla, California 92093 USA

MEMOIRS of the American Mathematical Society

SUBMISSION. This journal is designed particularly for long research papers (and groups of cognate papers) in pure and applied mathematics. The papers, in general, are longer than those in the TRANSACTIONS of the American Mathematical Society, with which it shares an editorial committee. Mathematical papers intended for publication in the Memoirs should be addressed to one of the editors:

Ordinary differential equations, partial differential equations and applied mathematics to ROGER D. NUSSBAUM, Department of Mathematics, Rutgers University, New Brunswick, NJ 08903

Harmonic analysis, representation theory and Lie theory to AVNER D. ASH, Department of Mathematics, The Ohio State University, 231 West 18th Avenue, Columbus, OH 43210

Abstract analysis to MASAMICHI TAKESAKI, Department of Mathematics, University of California, Los Angeles, CA 90024

Real and harmonic analysis to DAVID JERISON, Department of Mathematics, M.I.T., Rm 2–180, Cambridge, MA 02139

Algebra and algebraic geometry to JUDITH D. SALLY, Department of Mathematics, Northwestern University, Evanston, IL 60208

Geometric topology and general topology to JAMES W. CANNON, Department of Mathematics, Brigham Young University, Provo, UT 84602

Algebraic topology and differential topology to RALPH COHEN, Department of Mathematics, Stanford University, Stanford, CA 94305

Global analysis and differential geometry to JERRY L. KAZDAN, Department of Mathematics, University of Pennsylvania, E1, Philadelphia, PA 19104-6395

Probability and statistics to RICHARD DURRETT, Department of Mathematics, Cornell University, Ithaca, NY 14853-7901

Combinatorics and number theory to CARL POMERANCE, Department of Mathematics, University of Georgia, Athens, GA 30602

Logic, set theory, general topology and universal algebra to JAMES E. BAUMGARTNER, Department of Mathematics, Dartmouth College, Hanover, NH 03755

Algebraic number theory, analytic number theory and modular forms to AUDREY TERRAS, Department of Mathematics, University of California at San Diego, La Jolla, CA 92093

Complex analysis and nonlinear partial differential equations to SUN-YUNG A. CHANG, Department of Mathematics, University of California at Los Angeles, Los Angeles, CA 90024

All other communications to the editors should be addressed to the Managing Editor, DAVID J. SALTMAN, Department of Mathematics, University of Texas at Austin, Austin, TX 78713.

General instructions to authors for

PREPARING REPRODUCTION COPY FOR MEMOIRS

> **For more detailed instructions send for AMS booklet, "A Guide for Authors of Memoirs."**
> **Write to Editorial Offices, American Mathematical Society, P.O. Box 6248,**
> **Providence, R.I. 02940.**

MEMOIRS are printed by photo-offset from camera copy fully prepared by the author. This means that the finished book will look exactly like the copy submitted. Thus the author will want to use a good quality typewriter with a new, medium-inked black ribbon, and submit clean copy on the appropriate model paper.

Model Paper, provided at no cost by the AMS, is paper marked with blue lines that confine the copy to the appropriate size.

Special Characters may be filled in carefully freehand, using dense black ink, or **INSTANT** ("rub-on") **LETTERING** may be used. These may be available at a local art supply store.

Diagrams may be drawn in black ink either directly on the model sheet, or on a separate sheet and pasted with rubber cement into spaces left for them in the text. Ballpoint pen is not acceptable.

Page Headings (Running Heads) should be centered, in CAPITAL LETTERS (preferably), at the top of the page — just above the blue line and touching it.

LEFT-hand, EVEN-numbered pages should be headed with the AUTHOR'S NAME;

RIGHT-hand, ODD-numbered pages should be headed with the TITLE of the paper (in shortened form if necessary).

Exceptions: PAGE 1 and any other page that carries a display title require NO RUNNING HEADS.

Page Numbers should be at the top of the page, on the same line with the running heads.

LEFT-hand, EVEN numbers — flush with left margin;

RIGHT-hand, ODD numbers — flush with right margin.

Exceptions: PAGE 1 and any other page that carries a display title should have page number, centered below the text, on blue line provided.

FRONT MATTER PAGES should be numbered with Roman numerals (lower case), positioned below text in same manner as described above.

MEMOIRS FORMAT

> **It is suggested that the material be arranged in pages as indicated below.**
> **Note: Starred items (*) are requirements of publication.**

Front Matter (first pages in book, preceding main body of text).

Page i — *Title, *Author's name.

Page iii — Table of contents.

Page iv — *Abstract (at least 1 sentence and at most 300 words).

Key words and phrases, if desired. (A list which covers the content of the paper adequately enough to be useful for an information retrieval system.)

*_1991 Mathematics Subject Classification._ This classification represents the primary and secondary subjects of the paper, and the scheme can be found in Annual Subject Indexes of MATHEMATICAL REVIEWS beginnning in 1990.

Page 1 — Preface, introduction, or any other matter not belonging in body of text.

Footnotes: *Received by the editor date.
Support information — grants, credits, etc.

First Page Following Introduction – Chapter Title (dropped 1 inch from top line, and centered). Beginning of Text.

Last Page (at bottom) – Author's affiliation.